R00069 56487

Volume 4

Essays in
Physics

Volume 4
1972

Essays in Physics

Edited by

G. K. T. Conn and **G. N. Fowler**
Department of Physics
The University of Exeter
Exeter, U.K.

Academic Press
London and New York

ACADEMIC PRESS INC. (LONDON) LTD.
24/28 Oval Road,
London NW1

United States Edition published by
ACADEMIC PRESS INC.
111 Fifth Avenue
New York, New York 10003

Copyright © 1972 by
ACADEMIC PRESS INC. (LONDON) LTD.

All Rights Reserved
No part of this book may be reproduced in any form by photostat, microfilm, or any other means, without written permission from the publishers

Library of Congress Catalog Card Number: 74-117120
ISBN: 0-12-184804-3

PRINTED IN GREAT BRITAIN BY
ROYSTAN PRINTERS LIMITED
Spencer Court, 7 Chalcot Road
London NW1

List of contributors

MICHAEL E. FISHER, Baker Laboratory, Cornell University, Ithaca, New York, U.S.A. (p. 43)

G. N. FOWLER, Department of Physics, University of Exeter, Exeter, England (p. 1)

A. M. HOWATSON, Engineering Science Department, University of Oxford, Oxford, England (p. 91)

WALTER THIRRING, Institute of Theoretical Physics, University of Vienna, Vienna, Austria (p. 125)

Preface

The method of successive approximations is well-known in Physics. Perhaps in this, the fourth volume of Essays in Physics, we are nearer our target in the series. Articles which bring out the interactions between and among the fields of physics are valuable in themselves and provide a fresh stimulus. Weak and strong interactions affect fields of study as well as phenomena.

Thus recent measurements on K-mesons are used by Thirring to demonstrate that gravitational and inertial mass are—at least in this case— directly proportional, and that to an extraordinary precision. Fisher ranges over phase transitions in liquid–gas, magnetic and superconducting systems to clarify the concept of "order". Howatson illustrates the need to approach the physics of the arc from different directions; Fowler in the first of these essays displays the resemblance of atoms, nuclei and nucleons.

Who says that Physics, the precise science, is not both catholic and conjunctive?

G. K. T. CONN

G. N. FOWLER

Contents

LIST OF CONTRIBUTORS	v
PREFACE	vii
Electromagnetic Structure of Nucleons. By G. N. FOWLER	1
Phase Transitions Symmetry and Dimensionality. By MICHAEL E. FISHER	43
Arc Physics. By A. M. HOWATSON	91
Gravitation. By WALTER THIRRING	125

Electromagnetic Structure of Nucleons

G. N. FOWLER

University of Exeter, Department of Physics, Stocker Road, Exeter,

I. Introduction	1
II. The Stanford linear accelerator (Neal 1968)	3
III. Elastic scattering, basic formulae	4
(A) Electron–proton scattering	4
(B) Electron–neutron scattering	11
IV. Experimental results	14
V. Theory of the elastic form factors	16
(A) Meson theory	16
(B) Composite model	20
VI. Electron–proton inelastic scattering	26
VII. High energy pp elastic scattering and proton structure	36
VIII. Conclusions	38
Acknowledgements	39
Bibliography	39
References	39

I. Introduction

The general field with which this essay is concerned is currently one of the most exciting in high energy physics. It is linked directly with the classic work of Rutherford on the structure of the atom and so has an immediate appeal which few other topics in high energy physics now have. In addition, and thanks to the remarkable technical expertise of a number of laboratories, it is providing a stream of experimental results the significance of which is plainly immense for our understanding of particle structure.

At the present time it would be premature to suggest that definite and detailed conclusions concerning the structure of nucleons can be drawn from the data. Instead speculation is the order of the day and many different models claim to account for the results (and this not simply a consequence of large error bars).

The simplest model which might be introduced is of course one which treats the electron and proton as point charges and point magnetic moments. The scattering should then fall off with scattering angle according to Rutherford's law modified by the presence of a magnetic dipole, at least when relativistic effects are ignored. In fact the most cursory examination of the data shows that the scattering at large angles is far less than that predicted by this model. The implication of this is that the electron or the proton or both are not structureless points. From purely electromagnetic processes not involving strongly interacting particles we have good grounds for believing that the electron can be regarded for present purposes as a point particle. We are thus led to consider the nucleon as a particle with structure. Various possibilities spring to mind, probably the most natural is to consider the nucleon as a source of mesons which exist as a kind of cloud round the nucleon and which thus give the nucleon a finite size. Another possibility is that the nucleon is actually composed of other very strongly interacting constituents which have still to be isolated (the quarks?). Indeed if this kind of crude model is reliable we might expect the nucleon to behave in high energy inelastic collisions somewhat like a collection of independent entities, like the electrons in an atom or the nucleons in a nucleus. We shall see that this is indeed what might be happening.

In the following we shall discuss these scattering processes with particular reference to the composite model. This is of course to take a possibly narrow view but taking the data as a whole, this approach seems to be the most promising and it has an immediacy which can be very helpful in systematizing the rapidly growing quantity of data in the field. In any event it is arguably not the function of the contributors to these essays to be comprehensive but hopefully to bring out the key ideas in a reasonably accessible way.

We have also been selective in our description of the experimental procedures and have concentrated on the S.L.A.C. machine, because it has the highest energy presently available. The other major laboratories in the field are Stanford (Mk. III) (U.S.A.), Orsay (France), Cornell (U.S.A.), C.E.A. (U.S.A.), D.E.S.Y. (West Germany) and N.I.N.A. (U.K.).

We begin with a brief account of the essential features of the Stanford machine which has made such a significant contribution to the field, and then recapitulate the basic formulae used in the analysis of experiments on elastic scattering on protons and deuterons so as to extract the basic quantities which are to be compared with experiment. For a full understanding of these formulae the reader should have some knowledge of Maxwell's equations and relativistic quantum mechanics.

The experimental results are then described and their bearing on the formulae explained.

We then turn to a theoretical analysis of the results in terms of both the meson cloud and the composite pictures emphasizing the seminal pieces of information which help to distinguish between these models.

We then consider the experimental results on inelastic processes at very high energy and large angles and demonstrate the rather striking way in which the data suggest a composite model. Finally we conclude with a brief survey of other data bearing on the structure of nucleons arising from a study of high energy proton–proton elastic scattering.

II. The Stanford linear accelerator (Neal 1968)

In the S.L.A.C. machine, electrons (and positrons) are accelerated down a long evacuated tube by applying a radio frequency electric field. The most efficient way of doing this is to operate the accelerator tube as a wave guide and to arrange for the accelerating electrons to travel with the wave crests down the guide in much the same way as a surf rider. Since the phase velocity of the waves in the tube exceeds c it is necessary to load the guide by introducing discs with central holes to reduce the phase velocity to a value very close to c, which after a short time is essentially the electron velocity. Continued acceleration then serves in effect to increase the electron mass. However in order to produce the electrons of energies of 20 GeV or more, required to probe the structure of a nucleon, the tube length must be of the order of 3000 m. Obviously radio frequency energy is consumed as the electrons are accelerated down the pipe and the guide must be fed at regular intervals with synchronized, radio-frequency power. The density of electrons which can be accelerated simultaneously is limited by the mutual electrostatic repulsion which leads to an inability to focus the beam effectively, and also to loss of power through production by the electrons of resonating electric fields in the guide. In fact the peak beam current presently available is of the order 25–50 mA and if this were in fact the average beam current the average beam power would be ~ 100 MW. Assuming that, say, 10% of the input power is actually used to energize the beam this would require a power dissipation of 300 kW per metre of the accelerator. This is plainly quite unacceptable and in fact the beam of electrons is pulsed at 1 to 360 pulses per second, each pulse lasting 0.01–2.1 µs. In these circumstances the average beam current is 15–30 µA. The mean beam power is now ~ 600 kW implying a dissipation of 1.8 kW/m. This is removed by cooling water flowing at a rate of 13 gal/min maintaining the tube temperature at about 113° F. The electrons are supplied by an electron gun similar in principle to that used in a cathode ray tube, modulated to produce these electron pulses. To reduce loss of electrons, each pulse is split into bunches arranged so that each bunch may ride on the crest of a wave, and since the operating radio frequency is 2856 MHz with about 5×10^{11} electrons per pulse there are $\sim 10^8$ electrons per bunch, each bunch being about the size and shape of an aspirin tablet.

The accelerating tube is about 10 cm in diameter and, in spite of its length, guiding the beam down a tube 3000 m long is not as impossible as it might seem since to the electrons the tube is Lorentz contracted to about 1 m.

Focussing and steering magnets are provided to maintain the beam shape and direction; the earth's magnetic field must be compensated. The accelerator tube is constructed in 3 m sections and these are mounted in groups of 4 on a 12 m aluminium girder 60 cm in diameter. Inside the girder a laser beam is used to check the alignment.

The accelerating R.F. power is provided by 245 klystron tubes each of which feeds four 3 m sections of accelerator. Since the length of the electron pulse is approximately 480 m the klystrons are switched on and off with a slight delay between each so that the klystrons only operate for the requisite 2 µs timed to coincide with the arrival of the electron pulse. To accomplish this it is necessary to feed each klystron with a low power high frequency synchronizing signal, so that the oscillations of each klystron are in phase, and, in addition, a high power pulsed main supply so that it operates at full power whilst the electron pulse is present.

A positron beam can also be accelerated by the machine and 6 GeV electrons are used to produce them by bombardment of copper and tungsten targets through γ-ray production. The γ rays produced subsequently generate electron–positron pairs of which the positron component can be extracted by suitable deflecting fields. In this way positrons of ~ 12 GeV can be produced with mean currents of ~ 0.45 µA.

Observation and measurement of the scattered electrons requires a complex of instruments. Three magnetic spectrometers are used appropriate to momenta of 1.6, 8 and 20 GeV/c; these accept electrons scattered through various angles covering all possibilities. The momentum resolution is about 0.1% and the angular resolution about 0.0003 radians. To gain some idea of the size of these instruments it suffices to mention that the largest is 52 m long and weighs 1700 tons. A hydrogen (or deuterium) bubble chamber of sensitive volume 170 cubic feet is in use at S.L.A.C.

III. Elastic scattering, basic formulae

A. ELECTRON–PROTON SCATTERING

We are concerned in the main with the more easily accessible electron–proton scattering process. The electron–neutron results are then obtained from a parallel study of electron–deuteron scattering which involves further uncertainties arising from the participation of three bodies in the scattering event.

To begin with, consider the scattering of an electron by an extended charge distribution, for example an atomic nucleus, treated in lowest Born approximation. Using well known procedures of non-relativistic quantum theory, we have for the scattering amplitude,

$$f(\theta) = \frac{2\mu e}{4\pi\hbar^2} \int \exp(-i\mathbf{k}\cdot\mathbf{r}) \frac{\rho(\mathbf{r}')}{|\mathbf{r}-\mathbf{r}'|} \exp(i\mathbf{k}\cdot\mathbf{r}) \, d^3r \, d^3r',$$

where $\mathbf{k}$, $\mathbf{k}'$ are the wave numbers of the incoming and outgoing electron μ is the reduced mass of the colliding particles and $\rho(\mathbf{r})$ is the charge distribution satisfying

$$\int \rho(\mathbf{r}) \, d^3r = Ze.$$

Introducing the Fourier transform of the charge distribution

$$\rho(\mathbf{r}) = \frac{1}{(2\pi)^3} \int \exp(i\mathbf{k}\cdot\mathbf{r}) \, \tilde{\rho}(\mathbf{k}) \, d^3k,$$

we find

$$f(\theta) = \frac{2\mu e}{\hbar^2} \frac{\tilde{\rho}(\mathbf{k}-\mathbf{k}')}{|\mathbf{k}-\mathbf{k}'|^2}.$$

If $\rho(\mathbf{r})$ were a point charge distribution of magnitude Ze then the usual point charge scattering formula is immediately recovered. To include scattering by the nuclear magnetic moment, we must take account of the magnetic interaction with both the intrinsic and orbital magnetic moment of the electron. The contribution to the scattering produced by interaction with the nuclear magnetic moment distribution may be found using the associated vector potential $\mathbf{A}(\mathbf{r})$ given by

$$\mathbf{A}(\mathbf{r}) = \int \frac{\nabla \wedge \mathbf{M}(\mathbf{r}')}{|\mathbf{r}-\mathbf{r}'|} \, d^3r'.$$

Instead of the interaction

$$e \int \frac{\rho(\mathbf{r}')}{|\mathbf{r}-\mathbf{r}'|} \, d^3r',$$

one has $\mathbf{A}(\mathbf{r}) \cdot \mathbf{j}/c$ where $\mathbf{j}$ is the current produced by the incident electron.

The magnetic scattering amplitude is approximately

$$f_M(\theta) = \frac{2\mu e}{\hbar^2} \frac{(\mathbf{k}-\mathbf{k}') \wedge \tilde{\mathbf{M}}(\mathbf{k}-\mathbf{k}')}{|\mathbf{k}-\mathbf{k}'|^2} \cdot \mathbf{v}/c.$$

where $\tilde{\mathbf{M}}$ is the Fourier transform of the magnetic moment distribution and $\mathbf{v}$ the electron velocity. For a point dipole distribution $\tilde{\mathbf{M}}$ would simply be replaced by a constant vector, $\mathbf{m}$, the point dipole moment. The electron spin contribution can be included by considering the interaction between the magnetic field produced by the nucleus, $\mathbf{V} \wedge \mathbf{A}$, and the electron magnetic moment.

From this brief discussion it is clear that the electron scattering distribution is closely related to the Fourier transforms of the electric charge and magnetic moment distribution of the nucleus. This is, of course, a very general result valid—*mutatis mutandis*—for waves of many different kinds and in the present context it serves simply to focus attention on those properties of nucleons with which we shall chiefly be concerned without the added complications introduced by relativity. One further remark which is relevant concerns the difference between the angular distributions of the electric and the magnetic scattering. The electric distribution is more sharply peaked in the forward direction. This is a consequence of the less rapid fall off with distance of the electric field produced by the scattering centre compared with its dipole magnetic field. Thus small scattering angles produced by weaker long distance interactions are more likely in the former case.

Turning now to the relativistic treatment, the process is described by the Feynman diagram of Figure 1. Here p and p' refer to the incoming and outgoing electron, P and P' to the incoming and outgoing proton respectively, and $q_\mu = p_\mu - p_\mu'$ is the four momentum carried by the virtual photon represented by the line AB. (We use the metric $q^2 = \mathbf{q}^2 - q_0^2$ so that in the scattering process q_μ is space-like ($q^2 > 0$) and the units are such that $\hbar = c = 1$).

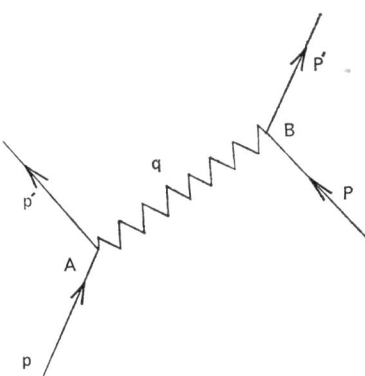

FIGURE 1. Feynman diagram representing ep scattering through the exchange of a virtual photon.

The vertices A and B are represented by the electron and proton current operators j_μ and J_μ respectively where

$$j_\mu = - ie\, \overline{U}_e(p')\gamma_\mu U_e(p),$$

and

$$J_\mu = ie\, \overline{U}_p(P')\, \{\gamma_\mu F_1(q^2) + iK\sigma_{\mu\nu}q_\nu F_2(q^2)\}\, U_p(P),$$

so that the diagram of Figure 3 simply describes the interaction between two electromagnetic currents of spin $\tfrac{1}{2}$ particles under relativistic conditions. Notice the extra factor of q_ν in the second term. This is part of the magnetic interaction, $\sigma_{\mu\nu}$ being the nucleon spin operator. U, $\overline{U}$, γ_μ, $\sigma_{\mu\nu}$ are the usual spinors and Dirac matrices (which we shall not need to use explicitly). K is the anomalous magnetic moment of the proton in nuclear magnetons and $F_1(q^2)$ and $F_2(q^2)$ are the elastic form factors we are principally interested in, normalized so that $F_1(0) = F_2(0) = 1$. These form factors F_1 and F_2 are called the Dirac and Pauli form factors for historical reasons. (If K were zero the proton would automatically have a magnetic moment of 1 nucleon magneton. The additional term in the electromagnetic interaction with form factor F_2 results in an additional, essentially arbitrary, magnetic moment and was first suggested by Pauli). If $F_1 = F_2 = 1$ for all q^2 the proton would be a point particle.

It is however more convenient to express the data in terms of two other form factors G_M and G_E which are related to F_1 and F_2 by

$$G_E(q^2) = F_1(q^2) - \tau K F_2(q^2), \tag{1}$$

$$G_M(q^2) = F_1(q^2) + K F_2(q^2), \tag{1'}$$

where

$$\tau = \frac{q^2}{4M^2}.$$

and M is the proton mass.

The differential scattering cross-section takes the form

$$\frac{d\sigma}{d\Omega} = \frac{\alpha^2 r_e^2 m^2}{4E^2 \sin^2(\theta/2)} \cdot \frac{E'}{E} \left\{ \frac{\cot^2(\theta/2)}{1+\tau} G_E^2(q^2) \right.$$

$$\left. + \tau G_M^2(q^2) \left[2 + \frac{\cot^2(\theta/2)}{1+\tau} \right] \right\}, \tag{2}$$

first given by Rosenbluth.

This may also be written in the form

$$\frac{d\sigma}{d\Omega} = \left(\frac{d\sigma}{d\Omega}\right)_M \left\{ \frac{G_E^2 + \tau G_M^2}{1+\tau} + 2\tau G_M^2 \tan^2(\theta/2) \right\}, \tag{3}$$

where

$$\left(\frac{d\sigma}{d\Omega}\right)_M = \left(\frac{\alpha r_e m}{2E}\right)^2 \frac{\cos^2 \theta}{\sin^4 (\theta/2)} \frac{1}{(1 + (2E/M) \sin^2 (\theta/2))}. \tag{3'}$$

Note that

$$q^2 = \frac{4E^2 \sin^2 (\theta/2)}{1 + (2E/M) \sin^2 (\theta/2)}.$$

Once again the magnetic term contains extra factors of q_v for the same reasons as before.

Here α, r_e and m are, respectively, the fine structure constant, the electron Compton wavelength and the electron mass. E and E' are the initial and final electron energies and θ is the scattering angle. The quantities G_E and G_M may in principle be found from experiment by plotting the quantity

$$\left(\frac{2E \sin (\theta/2)}{\alpha r_e m}\right)^2 \frac{E}{E'} \frac{d\sigma}{d\Omega} \quad \text{against} \quad \cot^2 (\theta/2)$$

for fixed q^2. The intercept at $\theta = 180°$ gives $G_M^2(q^2)$ and the slope then determines $G_E^2(q^2)$. Alternatively at each value of q^2, the cross-section is measured at the smallest angle allowed by the incident energy (since $q^2 \approx E^2\theta^2$). This gives $G_E^2 + \tau G_M^2$. Then at the same q^2, the counting rate is measured at the largest angle giving an acceptable counting rate, bearing in mind that the spectrometer solid angle may be fixed. In fact the form of the cross-section and the mode of analysis of the experimental results shows that the errors in F_1 and F_2 will generally be larger than those in G_E and G_M.

The physical interpretation of the form factors G_M and G_E may be understood by considering the scattering process in the Breit frame of reference in which the initial and final proton momenta are equal. This is reached from the laboratory frame, in which the proton is at rest, by a Lorentz transformation with velocity $\mathbf{q}/2M$. In this frame proton current conservation expressed by

$$\frac{\partial J_\mu}{\partial x_\mu} = 0$$

or in Fourier space

$$q_\mu \tilde{J}_\mu = 0,$$

becomes simply $\mathbf{q}_B \cdot \mathbf{J}_B = 0$, (since $q_0 = 0$ in the Breit frame). The current operator therefore consists of a scalar and a transverse vector which may be written

$$\tilde{J}_{OB} = e\, G_E(q^2)\chi_f{}^*\chi_i, \qquad (4)$$

$$\tilde{\mathbf{J}}_B = ie\, G_M(q^2)\chi_f \frac{\boldsymbol{\sigma}\wedge\mathbf{q}_B}{2M}\chi_i, \qquad (4')$$

where the χ_i, χ_f are spin wave functions and the σ_i are the Pauli matrices. In this frame G_E corresponds to the Fourier transform of the electric charge density and G_M to the transform of the magnetic moment density, hence their respective suffixes. In the non-relativistic case in which proton recoil is neglected, the Breit frame and laboratory frame coincide and the phsyical significance of G_E and G_M is clear. In the relativistic case this clear correspondence is no longer valid because the Breit frame varies with momentum transfer. This is of course a necessary consequence of the Lorentz covariance of the theory.

Another way of arriving at the Rosenbluth formula, which demonstrates its origin rather clearly, is by way of the process

$$e^+ + e^- \to p + \bar{p},$$

through a single intermediate virtual photon, which is illustrated in Figure 2. Considered as a function of q^2 the contribution of this diagram

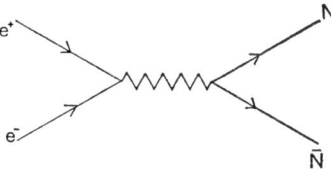

FIGURE 2. Feynman diagram representing the process $e^- + e^+ \to p + \bar{p}$.

is simply the analytic continuation of that from Figure 3 continued from $q^2 > 0$ to $q^2 < 0$. In particular, since the photon has spin 1, conservation of angular momentum requires that the initial and final total angular momentum must also be 1. When conservation of parity is taken into account the only states of e^-, e^+ (and $p, \bar{p}$) which are allowed are $^3S_1$ and $^3D_1$. The first gives an isotropic angular distribution, the second gives terms in $\cos^2 \phi$ where ϕ is the angle between the incoming and outgoing particles in the centre of mass system. Thus the cross-section is proportional to

$$\alpha(q^2) + \beta(q^2) \cos^2 \phi.$$

In fact for $p + \bar{p} \to e^- + e^+$

$$\left(\frac{d\sigma}{d\Omega}\right)_{G_M} = \frac{\alpha^2}{16M^2\tau(\tau(1+\tau))^{\frac{1}{2}}} \left\{ \tau G_M^2(q^2) - G_E^2(q^2) + \cos^2 \phi \left(\tau G_M^2(q^2) + G_E^2(q^2)\right) \right\}$$

where now $q^2 < -4M^2$.

When this is expressed in terms of Lorentz invariants and continued in q^2 to $q^2 > 0$ one recovers the corresponding quantity for electron–proton scattering. The continuation between ϕ and θ, the laboratory scattering angle of the e–p system, then turns out to be

$$\sin^2 \phi \to -\frac{\cot^2 \theta/2}{1+\tau},$$

and the Rosenbluth formula is quickly recovered apart from kinematic factors. Thus the $G_{E,M}$ representation affords in principle a check on the one exchange model depicted in Figure 1. Finally one constraint commonly imposed on G_E and G_M arises from the relations $(1, 1')$ between the G functions and the F's.

Expressed in terms of the G's we have for the F functions

$$K F_2(q^2) = \frac{G_M(q^2) - G_E(q^2)}{1+\tau}, \tag{5}$$

$$F_1(q^2) = \frac{G_E(q^2) + \tau G_M(q^2)}{1+\tau}. \tag{5'}$$

It follows that at $\tau = -1$ the F functions are infinite unless $G_M = G_E$ at this point. The value $\tau = -1$ corresponds to the threshold for the process $e^- e^+ \to p\bar{p}$ and, in physical terms, it seems reasonable to expect the F functions to be finite. In addition the angular distribution of the $e^- e^+$ process should be isotropic at threshold which again requires $G_M = G_E$ at $\tau = -1$.

ELECTROMAGNETIC STRUCTURE OF NUCLEONS 11

B. ELECTRON–NEUTRON SCATTERING

The principal source of information on the neutron form factor is the scattering of electrons by deuterons. Since this is a three body process for which exact solutions are not available, it is clear that additional uncertainty is involved in extracting the neutron form factor from the experimental data. The procedure for doing this has been described by a number of authors and we follow that given by Gourdin (1963, 1964, 1965). We may make contact with the discussion on the proton form factors by assuming, to begin with, that the collision takes place with each nucleon independently of the other. This is the so-called impulse approximation. If we ignore the internal relative motion within the deuteron then the deuteron Breit-frame is the same as that of the nucleons themselves and we may write down expressions for the electromagnetic current components similar to (5) and (5'). If in addition, we include a term linear in the momentum of the spectator nucleon we can make allowance for the omission of relative motion made earlier. Thus we have

$$\tilde{J}_{0B}^{(d)} = e\, G_{\text{ch}}(q^2) \chi_f^* \chi_i, \tag{6}$$

$$\tilde{J}_B^{(d)} = \frac{ie}{2M} \{G_{\text{mag}}(q^2) \chi_f^* \,\boldsymbol{\sigma} \wedge \mathbf{q}\chi_i - G_{\text{ch}}(q^2) \mathbf{P} \chi_f^* \chi_i\}. \tag{6'}$$

Here **q** is the three momentum transfer and in this frame is the difference between the nucleon momenta, **P** is the sum of the nucleon momenta. Since the deuteron is an isoscalar and since we are using the impulse approximation, only the isoscalar part of the form factors intervenes (see Section VI).

The final formula neglecting certain relativistic corrections is

$$\frac{d\sigma}{d\Omega} = \left(\frac{d\sigma}{d\Omega}\right)_M \left\{\frac{G_{\text{ch}}^{\,2}(q^2)}{1+\tau_d} + \frac{8}{9}\frac{\tau_d^{\,2}}{1+\tau_d} G_Q^{\,2}(q^2)\right.$$

$$\left. + \tfrac{2}{3}\tau_d(1+\tau_d)\left(\frac{1}{1+\tau_d} + 2\tan^2\theta/2\right) G_{\text{mag}}^2(q^2)\right\}. \tag{7}$$

In this expression,

$$G_{\text{ch}}^{\,2}(q^2) = 2\, G_{E,S}^{\,2}(q^2)\, C_E(q^2),$$

$$G_Q^{\,2}(q^2) = 2\, G_{E,S}^{\,2}(q^2)\, C_Q(q^2),$$

$$G_{\text{mag}}^2(q^2) = \frac{M_d}{M}[2\, G_{M,S}^{\,2}(q^2)\, C_S(q^2) + 2\, G_{E,S}^{\,2}(q^2)\, C_L(q^2)],$$

where
$$C_E = \int_0^\infty (U^2 + W^2) j_0\left(\frac{qr}{2}\right) dr,$$

$$C_Q = \frac{3\sqrt{2}}{2\tau_d} \int_0^\infty \left(UW - \frac{W^2}{2\sqrt{2}}\right) j_2\left(\frac{qr}{2}\right) dr,$$

$$C_s = \int_0^\infty \left(U^2 - \frac{W^2}{\sqrt{2}}\right) j_0\left(\frac{qr}{2}\right) dr + \frac{1}{2} \int_0^\infty \left(UW + \frac{W^2}{\sqrt{2}}\right) j_2\left(\frac{qr}{2}\right) dr.$$

U and W are wave functions representing the S- and D-state relative motion of the nucleons within the deuteron. The functions j_0 and j_2 are the usual spherical Bessel functions.

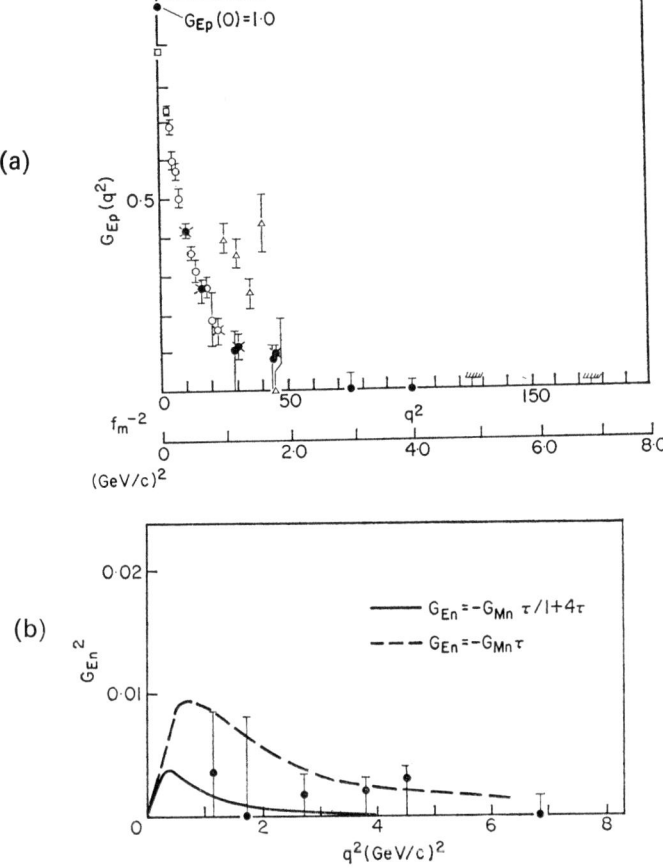

FIGURE 3(a). Electric form factor of proton as a function of q^2 Wilson (1967). (b) Electric form factor of neutron as a function of q^2 Budnitz et al. (1968).

The normalization of the G's is

$$G_{ch}(0) = 1, \quad G_Q(0) = M_d^2 Q, \quad G_{mag}(0) = 2M_d \mu_d,$$

where Q and μ_d are the quadrupole and magnetic moments of the deuteron respectively. Detailed knowledge of the deuteron wave function is clearly required if we are to extract information on the scalar part of the form factors and so isolate the neutron form factors themselves. It is therefore not surprising that the experimental errors indicated in Figure 3b are rather large.

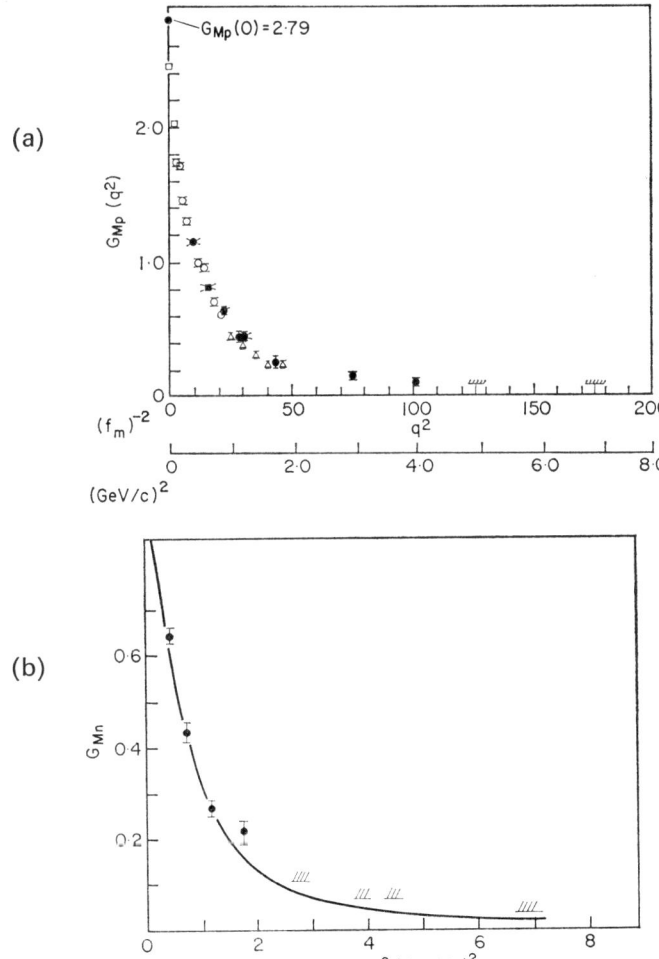

FIGURE 4(a). Magnetic form factor of proton as a function of q^2 Wilson (1967). (b) Magnetic form factor of neutron as a function of q^2 Budnitz et al. (1968).

In addition to electron–deuteron scattering, there are other experiments which bear on the neutron charge structure, in particular the scattering of thermal neutrons by elements of large Z. These give a value for the slope of the electric form factor of the neutron at $q^2 = 0$. The most recent experimental result is

$$\left(\frac{dG_{E,n}}{dq^2}\right)_{q^2=0} = 0.50 \pm 0.01 \; (\text{GeV}/c)^{-2},$$

which agrees with that from elastic electron deuteron scattering according to Rutherglen (1969).

IV. Experimental results

The experimental results shown in Figures 3, 4, 5 and 6 illustrate the manner in which the proton form factors differ from the point charge prediction—which would of course require $G_{E_p} = 1$ and $G_{M_p} = 2.79$, both for all q^2.

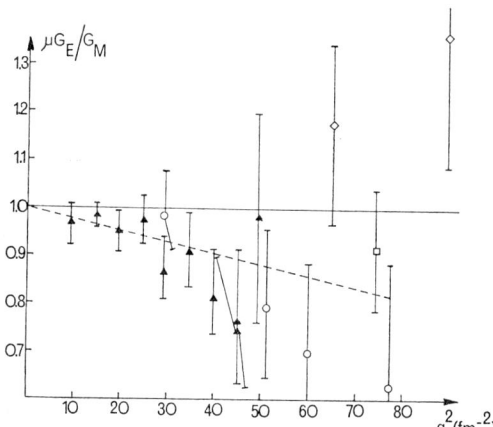

FIGURE 5. Ratio $\mu G_E/G_M$ as a function of q^2 where G_E and G_M are the electric and magnetic form factors of the proton and μ the proton magnetic moment.
▲ Experimental data of Ch. Berger et al. (1971).
○ Experimental data of W. Bartel et al. (1970).
□ Experimental data of W. Bartel et al. (1967).
◇ Experimental data of J. Litt et al. (1970).
The dashed line is given by $G_E/G_M = 1 - dq^2$, $d = (0.0023 \pm 0.0008)\text{fm}^2$.

Empirical features revealed by the data are first, the relation between the electric and magnetic form factors,

$$\frac{\mu_n G_{E,p}(q^2)}{G_{M,n}(q^2)} = \frac{\mu_p G_{E,p}(q^2)}{G_{M,p}(q^2)} \approx 1$$

where $\mu_{n,p}$ is the neutron/proton magnetic moment. This if often referred to as the scaling law, and is evidenced in Figure 5. Secondly, the empirical data are surprisingly well fitted by the simple formula

$$G_{E,p}(q^2) = \left[\left(1 + \frac{q^2}{0.71}\right)^2\right]^{-1}.$$

This is called the 'dipole' formula, $G_D(q^2)$. This feature of the data is shown for small and large q^2 in Figure 6a and b. Neither of these features is understood at the present time.

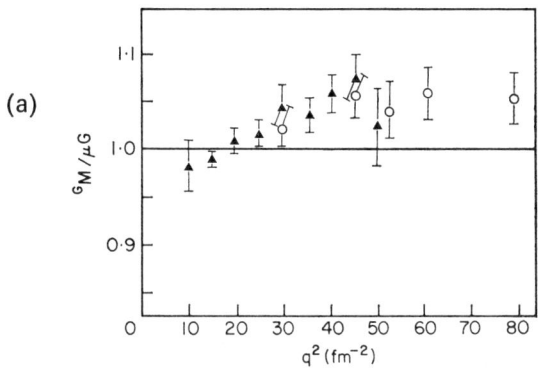

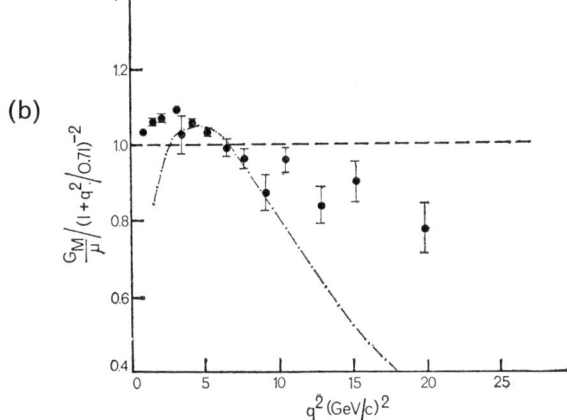

FIGURE 6(a). Ratio $G_M/\mu G_D$ as a function of q^2 for $q^2 < 3$ GeV/c^2, where G_M is the magnetic form factor of the proton and G_D the dipole form factor of Section IV.
▲ Experimental data of Ch. Berger et al. (1971).
○ Experimental data of W. Bartel et al. (1970).
(b) The ratio $G_M/\mu G_D$ as a function of q^2 for $q^2 > 3$ GeV/c^2. D. H. Coward et al. (1968). Also shown is the fit given by the theory of Wu and Yang (1965). (See Section VII).

V. Theory of the elastic form factors

A. MESON THEORY

As we have seen the experimental results indicate gross departures from the predictions of straightforward perturbation theory, that is to say the functions $G_M(q^2)$ and $G_E(q^2)$ are very far from being constant. The first and most obvious reason for this may be that the electromagnetic interaction with nucleons may be modified by the mesonic fields of the nucleon. In other words the interaction between the electromagnetic field and the nucleons takes place in the first instance, especially for not too large q^2, through interaction with mesons. In terms of the relevant field operators this means that the electromagnetic interaction Hamiltonian density takes the form

$$H_1(x) = e A_\mu(x) j_\mu^e(x) + e A_\mu(x) j_\mu^M(x) + f F_\mu(\phi) J_\mu^N(x). \tag{8}$$

The first two terms represent coupling of the electromagnetic field with electrons and mesons respectively and the last term describes coupling of the mesons with nucleons. The function $F_\mu(\phi)$ is a four vector function of the meson field which remains to be specified and J_μ^N is the nucleon electromagnetic current operator; for the moment we do not distinguish between electric and magnetic interactions. We shall use the equation (8) merely as a guide to the general form which G_M and G_E might take.

To begin with we assume that the mesonic current takes the usual form

$$j_{\mu V}^M = i\left(\phi \frac{\partial}{\partial x_\mu} \phi^\dagger - \phi^\dagger \frac{\partial}{\partial x_\mu} \phi\right), \tag{9}$$

where the suffix V refers to the isovector property of the mesonic current given by (9), the function ϕ being an isovector. (See Eden (1970) for a discussion of isospin.) It is important to recognize however that the nucleon charge may be expressed in terms of isospin operators by

$$Q = \tfrac{1}{2}(\tau_3 + 1),$$

where τ_3 is the 3 component of the nucleon isospin operator. This implies that the nucleon electromagnet operator expressed in terms of its isospin characteristics is a linear combination of isovector and isoscalar. Since strong interactions conserve isospin, the expression (9) for the mesonic current is not enough to describe the electromagnetic structure of the nucleon, we must have an isoscalar term in the current operator. Expressed in terms of particles the operator (9) either leaves the number of mesons in the field unchanged or changes it by two with total charge zero. One may look at the coupling of photons with meson pairs in more detail by noting that the pions being bosons can only exist in totally symmetric states.

A pair of mesons must, in order to couple to a photon, have total angular momentum $L = 1$ (in order to conserve angular momentum) and so be in an odd space state. This requires that they be in a state of odd isospin which in turn requires that the total isospin $T = 1$. Finally charge conservation in the coupling to photons requires $T_3 = 0$. To produce an $L = 1$ isoscalar combination, we must have at least 3 mesons which implies coupling of the electromagnetic field to the meson field involving the third power of ϕ. The function $F_\mu(\phi)$ in the meson nucleon coupling must therefore be chosen to emit or absorb isovector or isoscalar spin 1 (i.e. spatial vector) mesons. It will be clear that single mesons do not enter into the discussion but only states with an even number or an odd number >1. We shall restrict these numbers to 2 and 3.

The simplest thing to do now is to assume that we are dealing with two quasi elementary vector meson fields of isovector and isoscalar type respectively which couple to nucleons directly through terms of the type

$$g_V B_\mu^M J_{\mu,v}^N + g_S C_\mu^M J_{\mu S}^N,$$

where $J_{\mu,v}^N$ and $J_{\mu,s}^N$ are the isovector and isoscalar parts of the nucleon electromagnetic current. The coupling of these mesons to the electromagnetic field is assumed to be through terms of the type

$$f_V A_\mu B_\mu^M + f_S A_\mu C_\mu^M.$$

Another possibility for coupling the vector mesons to photons could be used, which moreover has the advantage of gauge invariance, for example

$$g' G_{\mu\nu} F_{\mu\nu}$$

where

$$G_{\mu\nu} = \frac{\partial B_\mu^M}{\partial x_\mu} - \frac{\partial B_\mu^M}{\partial x_\nu}$$

and $F_{\mu\nu}$ is the corresponding quantity for the A_μ. For our purpose it suffices to choose the simpler version. This vector meson model was first put forward by Sakurai (1960) and subsequently discussed in detail by Kroll et al., (1967).

The field equations for the vector potentials are

$$\frac{\partial^2 A_\nu}{\partial x_\mu^2} = f_V B_\nu^M + f_s C_\nu^M + e j_\nu^e, \qquad (10)$$

and the field equations for the vector mesons are

$$\frac{\partial^2 B_\nu^M}{\partial x_\mu^2} - \mu_V^2 B_\nu^M = f_V A_\nu + g_V J_{\nu,v}^N, \qquad (11)$$

and
$$\frac{\partial^2 C_\nu^M}{\partial x_\mu^2} - \mu_S^2 C_\nu^M = f_S A_\nu + g_S J_{\nu S}^N \tag{12}$$

using
$$\frac{\partial A_\mu}{\partial x_\mu} = \frac{\partial C_\mu^M}{\partial x_\mu} = \frac{\partial B_\mu^M}{\partial x_\mu} = 0.$$

Taking the Fourier transforms of (10), (11) and (12) we find that

$$\tilde{A}_V(q) = -\frac{e\tilde{j}_\nu^e(q) - f_V \tilde{B}_\nu^M(q) - f_S \tilde{C}_\nu^M(q)}{q^2}, \tag{13}$$

$$\tilde{B}_\nu^M(q) = -\frac{g_V \tilde{J}_{\nu,V}(q)}{q^2 + \mu_V^2} - \frac{f_V \tilde{A}_\nu(q)}{q^2 + \mu_V^2}, \tag{14}$$

$$\tilde{C}_\nu^M(q) = -\frac{g_S \tilde{J}_{\nu,S}(q)}{q^2 + \mu_S^2} - \frac{f_S \tilde{A}_\nu(q)}{q^2 + \mu_S^2}. \tag{15}$$

If we assume that the only source of electromagnetic field is the incident electron and the only source of mesons is the electromagnetic field so produced, we may replace (13), (14) and (15) by

$$\tilde{A}_V(q) = -\frac{e\tilde{j}_\nu^e(q)}{q^2}, \tag{16}$$

$$\tilde{B}_\nu^M(q) = -\frac{f_V \tilde{A}_\nu(q)}{q^2 + \mu_V^2} = +\frac{f_V e\tilde{j}_\nu^e(q)}{q^2 + \mu_V^2} \cdot \frac{1}{q^2}, \tag{17}$$

$$\tilde{C}_\nu^M(q) = -\frac{f_S \tilde{A}_\nu(q)}{q^2 + \mu_S^2} = +\frac{f_S e\tilde{j}_\nu^e(q)}{q^2 + \mu_S^2} \cdot \frac{1}{q^2}. \tag{18}$$

If we now substitute these expressions into the interaction energy we find a direct current interaction of the type

$$\tilde{H}_I(q) = \frac{g_V f_V e\tilde{j}_\nu^e(q) \cdot \tilde{J}_\nu^N(q)}{q^2(q^2 + \mu_V^2)} + \frac{g_S f_S e\tilde{j}_\nu^e(q) \cdot \tilde{J}_{\nu S}^N(q)}{q^2(q^2 + \mu_S^2)}. \tag{19}$$

We could of course have taken the nucleon as a source of mesons and worked backwards with the same result. It should be clear that the form factor which we find in this way is

$$G_{E,M,V} = \frac{\text{Const}}{q^2 + \mu_V^2},$$

$$G_{E,M,S} = \frac{\text{Const}}{q^2 + \mu_S^2}.$$

The form factors of the protons and the neutron are related to G_S and G_V by

$$G_V = \tfrac{1}{2}(G_P - G_N), \qquad G_S = \tfrac{1}{2}(G_P + G_N),$$

dropping for the moment the suffix E, M. There is no reason to suppose that the constants g_S and g_V are the same for the electric and magnetic couplings so that in total we have to consider 4 distinct quantities G_{EV} and G_{ES}, G_{EM} and G_{ES}.

It is interesting to note that the first predictions of the existence of vector mesons of the type described above came from calculations on nucleon form factors by Frazer and Fulco (1960) who were concerned with the isovector case. Indeed since that time a number of short lived vector mesons of isovector and isoscalar type have been discovered and initially hopes ran high that they would be the key to the understanding of the nucleon form factors. The conventional parameterization which arises in this way has the general form

$$G_{E,M} = \sum_\alpha \frac{g_{E,M,\alpha}}{q^2 + \mu_\alpha^2}, \tag{20}$$

where the summation over α refers to a sum over the possible vector mesons, isovector or isoscalar. Provided $g_{E,M}$ falls off sufficiently fast we may write

$$G_{E,M} = \frac{1}{\pi} \int_{\mu_{\min}}^{\infty} \frac{g_{E,M}(\mu^2)\, d\mu^2}{q^2 + \mu^2}. \tag{20'}$$

Indeed it may be shown that the form (20) follows from much more general assumptions and may be analytically continued in q^2 to $q^2 < 0$. The form factors then possess an imaginary part directly related to nucleon–antinucleon annihilation processes which we have already referred to briefly. The state of the art up to this point is reviewed by Hand et al., (1963) and Wilson and Levinger (1964).

Of course it is evident from the interaction used here that the possibility that the nucleon might interact directly with the photon, without the intermediary meson field, has been explicitly neglected. Such a term may be included by letting f_V and μ_V^2 tend to infinity in such a way that $f_V/\mu_V^2 \to$ constant; this would correspond to zero propagation distance for the meson in the limit and would simulate direct nucleon–proton interaction. The result would be to add to G_E and G_M a constant term, so that

$$\lim_{q^2 \to \infty} G_E = C_1 \quad \text{and} \quad \lim_{q^2 \to \infty} G_M = C_2,$$

where C_1 and C_2 are constants. According to a conjecture of Sachs (1962) $C_1 = Z_2 Q$ and $C_2 = Z_2 Q/2M$, where Q is the charge corresponding to this direct interaction and is 0 or 1 in units of the proton charge and $Q/2M$ the corresponding magnetic moment (in sin units). The quantity Z_2 could therefore be described as the probability that the electromagnetic field will encounter a nucleon stripped, as it were, of its mesonic field, that is to say, a bare nucleon. We shall encounter this quantity again later. Suffice it to say at this point that there is no evidence for a contribution of this kind, that is to say G_E and G_M apparently vanish in the limit $q^2 \to \infty$. The implication of this is that $Z_2 = 0$ and there is no elementary nucleon.

In fact in more recent times it has become clear, as has been mentioned in Section IV, that the form factors are best fitted by a form

$$G = \frac{\text{Const}}{(q^2 + m^2)^2},$$

the so-called dipole form (the name arises from the fact that in the $q^2 < 0$ region G has a double pole). This is of course quite inconsistent with (20) and indeed requires that

$$\int g(\mu^2)\,d\mu^2 = 0.$$

B. COMPOSITE MODEL

At the present time no satisfactory account of the elastic form factors exists and probably the most significant piece of information comes from inelastic electron–proton scattering with large four momentum transfer and inelasticity. The data for these events are strikingly consistent with a nucleon consisting of a number of subunits which interact quasi-elastically with the incident electron. We discuss this further in Section VII, but for the moment and in view of the conclusion concerning Z_2 mentioned above, it seems reasonable to examine the elastic data in terms of a composite model of the nucleon.

To see how the scattering by a composite nucleon might be expected to differ from that from an elementary, or non-composite, nucleon we discuss the scattering of a non-relativistic electron by a well-known composite system, the hydrogen atom. The elastic scattering amplitude is given to lowest order in the charge by

$$f(\mathbf{q}) = \frac{e^2}{\mathbf{q}^2} \int |\psi(r)|^2 \exp(i\mathbf{q}\cdot\mathbf{r})\,d^3r \equiv \frac{e^2}{\mathbf{q}^2} F(\mathbf{q}),$$

where $\mathbf{q}$ is the three momentum transfer and

$$\psi(\mathbf{r}) = N e^{-\alpha r}, \quad \text{with} \quad N = \left(\frac{\alpha^3}{\pi}\right) \quad \text{and} \quad \alpha = \frac{me^2}{\hbar^2}.$$

ψ is the ground state wave function of the hydrogen atom and $F(\mathbf{q})$ is the elastic atomic form-factor. The normalization of the wave function guarantees that $F(0) = 1$ (but see below). The result is

$$f(\mathbf{q}) = \frac{e^2}{\mathbf{q}^2} \cdot \frac{4\pi N^2 \cdot 4\alpha}{(\mathbf{q}^2 + 4\alpha^2)^2},$$

and the form factor $F(\mathbf{q})$ is

$$F(\mathbf{q}) = \frac{16\pi\alpha N^2}{(\mathbf{q}^2 + (2\alpha)^2)^2}.$$

Had the scattering centre been a point charge i.e. $\psi(\mathbf{r}) \to \delta^3(\mathbf{r})$, then

$$f(\mathbf{q}) = \frac{e^2}{\mathbf{q}^2},$$

and

$$F(\mathbf{q}) \equiv 1,$$

as we should expect.

To make contact with the nucleon case we represent the scattering by a diagram as in Figure 7 in which the incident electron sees the atomic electron only. Of course for very small $\mathbf{q}$ the scattering amplitude becomes very large in this case and the charge on the nucleus must be taken into account. In this approximation the normalization $F(0) = 1$ reflects the assumption that the incident particle only sees the atomic electron.

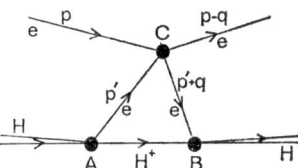

FIGURE 7. Triangle diagram representing the scattering of an electron by a composite system.

The diagram of Figure 7 can be brought in closer correspondence with $f(\mathbf{q})$ by introducing the Fourier transform $\phi(\mathbf{p})$ of the wave function $\psi(\mathbf{r})$ into the equation for $f(\mathbf{q})$ with

$$\psi(\mathbf{r}) = \frac{1}{(2\pi)^3} \int \exp(-i\mathbf{p}\cdot\mathbf{r})\, \phi(\mathbf{p})\, d^3\mathbf{p},$$

so that
$$\phi(\mathbf{p}) = \frac{N\alpha}{\pi^2(\alpha^2 + \mathbf{p}^2)^2}.$$

We have
$$f(\mathbf{q}) = \frac{e^2}{\mathbf{q}^2} \frac{N^2\alpha^2}{\pi^4(2\pi)^3} \int \frac{d^3\mathbf{p}'}{((\mathbf{p}' + \mathbf{q})^2 + \alpha^2)^2} \frac{1}{(\mathbf{p}'^2 + \alpha^2)^2},$$

$$\equiv \frac{e^2}{\mathbf{q}^2} \int \frac{d^3\mathbf{p}'}{(2\pi)^3} \frac{\Gamma(\mathbf{p}' + \mathbf{q})}{((\mathbf{p}' + \mathbf{q})^2 + \alpha^2)} \frac{\Gamma(\mathbf{p}')}{(\mathbf{p}'^2 + \alpha^2)}$$

with
$$\Gamma(p') \equiv \frac{N\alpha}{(\mathbf{p}'^2 + \alpha^2)\pi^2}$$

and
$$F(\mathbf{q}) = \int \frac{d^3\mathbf{p}'}{(2\pi)^3} \frac{\Gamma(\mathbf{p}' + \mathbf{q})}{((\mathbf{p}' + \mathbf{q})^2 + \alpha^2)} \frac{\Gamma(\mathbf{p}')}{(\mathbf{p}'^2 + \alpha^2)}.$$

Considering Figure 7, the vertices A and B are now represented by vertex functions $\Gamma(\mathbf{p}')$ and $\Gamma(\mathbf{p}' + \mathbf{q})$ respectively and vertex C by the scattering amplitude $e^2/\mathbf{q}^2$. The two lines (or propagators) representing the initial and final atomic electron are then given by the factors $1/(\mathbf{p}'^2 + \alpha^2)$ and $1/[(\mathbf{p}' + \mathbf{q})^2 + \alpha^2]$. (The nucleon line is represented by a constant in our infinite mass approximation). In our units $\alpha^2 = 2m E_B$ where E_B is the binding energy of the ground state. The vertex function Γ is then

$$\Gamma(\mathbf{p}) = (\mathbf{p}^2 + 2m E_B)\phi(\mathbf{p})$$

$$= \frac{1}{(2\pi)^3} \int V(\mathbf{r})\psi(\mathbf{r}) \exp(i\mathbf{p} \cdot \mathbf{r}) d^3r = \langle \mathbf{p}|V|\psi \rangle,$$

which follows from the Schrödinger equation, where V is the potential energy. Thus if
$$H_0 = \frac{\mathbf{p}^2}{2m},$$

we have
$$(E_B - H_0)|\psi_B\rangle = V|\psi_B\rangle$$

and
$$\Gamma(\mathbf{p}) = \left\langle \mathbf{p} \middle| V \frac{1}{E_B - H_0} V \middle| \psi_B \right\rangle$$

or
$$\Gamma(\mathbf{p}) = \int \frac{d^3\mathbf{p}'}{(2\pi)^3} \frac{\langle \mathbf{p}|V|\mathbf{p}'\rangle \Gamma(\mathbf{p}')}{E_B - \mathbf{p}'^2/2m}. \qquad (21)$$

So far we have considered the scattering by what is indubitably a composite particle. If we now wish to apply (21) to the nucleon we must modify it to allow for the possibility that the nucleon might after all not be composite. This means that Figure 7 might represent an incident nucleon which emits a virtual meson, propagates *as a nucleon*, and then reabsorbs the meson. Such a process might reasonably be expected to describe the charge and magnetic moment distribution of a nucleon in a perturbation approximation at least. This implies that there exists a state $|\psi_{B_0}\rangle$ corresponding to the elementary nucleon which satisfies

$$H_0 | \psi_{B_0} \rangle = E_{B_0} | \psi_{B_0} \rangle.$$

The effect of the interaction V is simply to change the nucleon mass from the bare mass to the observed value. This in turn means that the set of continuum states introduced in (21) is incomplete. We then allow for the state $|\psi_{B_0}\rangle$ and the result is

$$\Gamma(\mathbf{p}) = \sqrt{Z}\,\Gamma_0(\mathbf{p}) + \int \frac{d^3\mathbf{p}'}{(2\pi)^3} \frac{\langle \mathbf{p}|V|\mathbf{p}'\rangle \Gamma(\mathbf{p}')}{E_B - \mathbf{p}'^2/2m} \tag{22}$$

where

$$\sqrt{Z} = \langle \psi_{B_0} | \psi_B \rangle,$$

and

$$\Gamma_0(\mathbf{p}) = \langle \mathbf{p}|V| \psi_{B_0} \rangle.$$

The quantity Z is a renormalization constant which we identify with Z_2, encountered earlier. Evidently if $Z_2 = 0$ the nucleon is composite. On the other hand if $Z_2 \neq 0$ the first term in (22) might be expected to contain a point like core term which would imply $\Gamma_0(\mathbf{p}) \to$ constant as $\mathbf{p} \to \infty$. In this limit, $\Gamma(\mathbf{p})$ also tends to a constant value. The form-factor is then,

$$\lim_{q^2 \to \infty} F(\mathbf{q}) = \int \frac{d^3\mathbf{p}'}{(2\pi)^3} \frac{1}{((\mathbf{p}'+\mathbf{q})^2 + \alpha^2)} \frac{1}{(\mathbf{p}'^2 + \alpha^2)}$$

$$= \int \frac{e^{-2\alpha r}}{r^2} \exp(i\mathbf{q}\cdot\mathbf{r})\, d^3\mathbf{r},$$

$$= \frac{4}{|\mathbf{q}|} \int e^{-2\alpha r} \frac{\sin qr}{r}\, dr$$

$$= \frac{4}{|\mathbf{q}|} \left[-\tan^{-1}\frac{2\alpha}{q} + \frac{\pi}{2} \right].$$

Thus

$$\lim_{q^2 \to \infty} F(\mathbf{q}) = \frac{1}{|\mathbf{q}|},$$

in marked contrast with its $1/\mathbf{q}^4$ behaviour when $Z = 0$. It should be emphasized once again that we have still neglected the direct interaction term.

The conclusion from this discussion is that at least in this non-relativistic model the composite nucleon has a more rapidly decreasing form factor than the elementary nucleon. Such a treatment is at best illustrative of what might happen in a relativistic theory but more detailed consideration of the relativistic case shows that at least in the triangle approximation of Figure 7, the form factors F_1 and F_2 behave according to

$$F_1(\mathbf{q}) \approx \frac{(\ln \mathbf{q}^2)}{\mathbf{q}^4}$$

and

$$F_2(\mathbf{q}) \approx \frac{1}{\mathbf{q}^4},$$

taking account of nucleon spin and assuming that the nucleon is composite and appears as a bound state in the $P_{\frac{1}{2}}$ channel of pion–nucleon scattering. This compares with $1/\mathbf{q}^2$ behavior when $Z_2 \neq 0$.

In the relativistic case one finds that besides the quantity Z_2 other renormalization constants appear, in particular a coupling renormalization constant Z_1 associated with the existence of an elementary nucleon–meson interaction. It is this quantity which is associated with the possibility $\Gamma(\mathbf{p}) \to \Gamma_0(\mathbf{p}) \to$ constant as $\mathbf{p} \to \infty$. (For a fuller account of the ideas of renormalization theory including the introduction of the renormalization constants Z_1 and Z_2 see Pipkin, 1970). Hence one concludes that if $F_{1,2}(\mathbf{q}) \approx 1/\mathbf{q}^4$, $Z_1 = 0$. In fact the quantity $\langle \psi_{B_0} | \psi_B \rangle$, the wave function renormalization constant Z_2, must also vanish if the nucleon is composite, as one might expect.

Evidence from inelastic scattering which we discuss below does appear to substantiate the conclusion that $Z_2 \approx 0$. The argument is of course based on the assumption that the usual axioms of a renormalizable field theory are applicable to the nucleon and its constituents which is of course open to question.

It is perhaps worth summarizing at this point the arguments relating to the composite nucleon considering only the evidence from elastic scattering data.

In the first place it can be shown, using general arguments based on dispersion relations and assuming the validity of certain limiting processes, that

$$\lim_{q^2 \to \infty} F_1(\mathbf{q}^2) = Z_1^{(s)}.$$

$Z_1^{(s)}$ is a charge renormalization constant arising from the strong interactions in which the proton participates. This is not quite the same as

$$\lim_{q^2 \to \infty} G_E(\mathbf{q}^2) = Z_1^{(s)}.$$

For the latter to hold we must have

$$\lim_{q^2 \to \infty} \mathbf{q}^2 F_2(\mathbf{q}^2) = 0$$

which follows from the second of Sachs' conjectures $G_M(\mathbf{q}^2) \to Z_2 Q/2M$. Assuming this to be true, then from gauge invariance (charge conservation) we have $Z_1^{(s)} = Z_2$ where Z_2 is the wave function renormalization constant introduced earlier. However according to experiment

$$\lim_{q^2 \to \infty} G_E(\mathbf{q}^2) = 0,$$

so that $Z_2 = 0$ and there is no elementary nucleon. Furthermore according to the argument based on the triangle diagram of Figure 7 the presence of an elementary nucleon, which would correspond in this diagram to the presence of a direct nucleon–meson coupling at vertices A and B, would give

$$\lim_{q^2 \to \infty} \mathbf{q}^2 G_E(\mathbf{q}^2) = \text{const.}$$

As we have seen the experimental results indicate

$$\lim_{q^2 \to \infty} \mathbf{q}^2 G_E(\mathbf{q}^2) = 0,$$

so that there is no direct nucleon–meson coupling either. Thus $\tilde{Z}_1 = 0$, where $\tilde{Z}_1$ refers to the nucleon–meson coupling renormalization constant. According to Cornwall and Levy (1969) $\tilde{Z}_1 = 0$ also implies $Z_2 = 0$ when strong interactions only are considered. There is no question of gauge invariance here. Thus once again we arrive at the condition that there is no elementary nucleon. The interesting remark which appears to follow from this is that

$$\lim_{q^2 \to \infty} G_E(\mathbf{q}^2) = 0$$

requires in addition that

$$\lim_{\mathbf{q}^2 \to \infty} \mathbf{q}^2 G_E(\mathbf{q}^2) = 0$$

at least if all the models of composite nucleons so far considered bear any relation to reality.

VI. Electron–proton inelastic scattering

In the same way as for elastic scattering we may describe the inelastic scattering in terms of a diagram as in Figure 8. The final state of the nucleon will be, in general, a state which contains other particles—for example π mesons—restricted only by the requirement that it have baryon number $+1$ and charge $+e$.

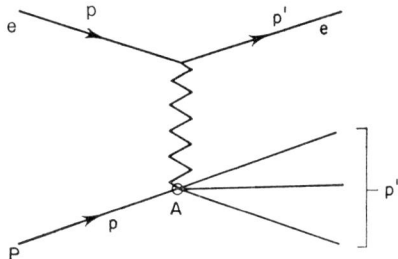

FIGURE 8. Feynman diagram representing inelastic ep scattering producing an arbitrary final nucleon state through the exchange of a virtual photon.

The analysis may be carried out in a manner similar to that for inelastic scattering by a hydrogen atom with the result that a vertex factor appears describing the process

$$\gamma + p \to p + \chi,$$

corresponding to vertex A where χ represents the other particles produced and γ is a virtual photon. This vertex factor evidently corresponds to photon absorption at A and indeed the only differences which arise between this and the case of real photons are in the selection rules which govern the transition at A: virtual photons may be longitudinally as well as transversely polarized and the mass of the virtual photon is not zero.

The differential cross-section takes the form

$$\frac{\partial^2 \sigma}{\partial q^2 \partial v} = \frac{\alpha}{2\pi} \frac{1}{E^2} \frac{K}{q^4} \left\{ q^2 \sigma_T(q^2, v) \right.$$

$$\left. + \frac{q^2}{\mathbf{q}^2} [\sigma_T(q^2, v) + \sigma_L(q^2, v)] (2EE' - q^2/2) \right\}. \quad (23)$$

Here $\sigma_T(q^2, v)$ and $\sigma_L(q^2, v)$ are the cross-sections for transverse and longitudinal virtual photon absorption with mass $-q^2$, α is the fine structure constant, v is the energy transfer, E, E' are the initial and final electron energies and K the energy in the laboratory system which a real photon would need to produce the final state $p + \chi$. For real photons $\sigma_L \equiv 0$ and $\sigma_T(0, K)$ is the real photon cross-section.

Early work by George and Evans (1950) on muon–nucleon inelastic scattering at high energies using the high energy muons found in cosmic rays gave a value of the order of 10^{-28} cm^2 for the photo–nucleon cross-section, $\sigma_T(0, K)$. This was interpreted as being resonance production of pions but the same cross-section was used successfully in most subsequent cosmic ray work at much higher energies. This led to the suggestion by Fowler and Wolfendale (1957) that smaller regions of the nucleon must be responsible for the high energy muon events. Recent work in this field is discussed by Cassidy (1971). Since that time the development of accelerators has led, as in so many other problems in high energy physics, to a vast increase in our understanding of inelastic events involving electrons and protons.

In treating the data, particularly since the work of Bjorken to which we now turn, a description of the differential cross-section in terms of other new parameters has been shown to be more useful. Two invariant functions $W_{1,2}(q^2, v)$ are introduced related to σ_T and σ_L by

$$W_1 = K \frac{\sigma_T(q^2, v)}{4\pi^2 \alpha},$$

and

$$\frac{W_1}{W_2} = \left(1 + \frac{v^2}{q^2}\right) \frac{\sigma_T}{\sigma_T + \sigma_L}.$$

P is the nucleon four momentum,

$$v = -\frac{q \cdot P}{M} = E - E',$$

the energy transfer in the laboratory system, as before, and

$$q^2 = 4EE' \sin^2(\theta/2),$$

evaluated in the laboratory system. Also useful in interpreting the data is W, the so-called missing mass. This is the energy in its centre of mass of the target nucleon and the virtual photon. This is given by

$$W^2 = M^2 - q^2 + 2Mv.$$

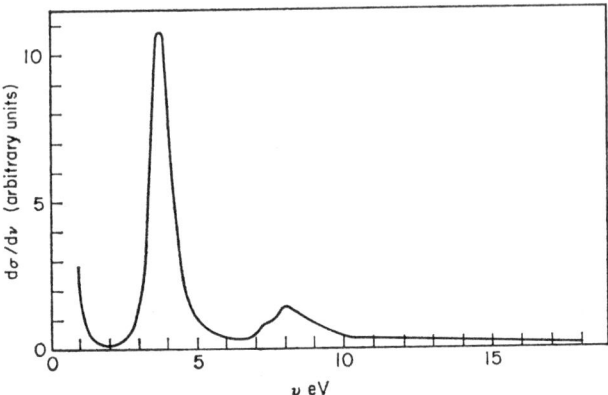

FIGURE 9. The energy loss spectrum $\partial\sigma/\partial v$ (arbitrary units) as a function of v for 25 keV electrons scattered inelastically by silver atoms. (H. Boersch et al., 1962.)

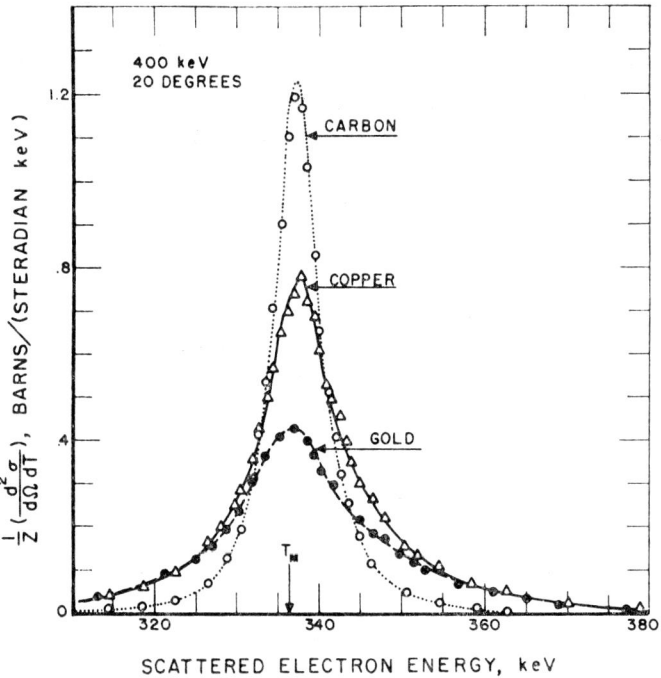

FIGURE 10. The energy loss spectrum $(1/Z)(\partial^2\sigma/\partial\Omega\partial T)$ as a function of final electron kinetic energy for 400 keV electrons, scattered inelastically by various elements in the form of thin films through an angle of 20°. The kinetic energy of the scattered electron corresponding to elastic scattering by an atomic electron initially at rest is indicated by the arrow labelled T_M. (G. Missoui et al., 1970.)

The differential cross-section then takes the form

$$\frac{\partial^2 \sigma}{\partial \Omega \partial v} = \frac{\alpha^2}{4E^2 \sin^4 (\theta/2)} \left(W_2 \cos^2 (\theta/2) + 2W_1 \sin^2 (\theta/2) \right)$$

when the electron mass is neglected. The utility of the functions W_1 and W_2 arises from the fact that in the limit $q^2 \to +\infty$,

$$vW_2 \to F_1 \left(\frac{q^2}{Mv} \right),$$

$$W_1 \to F_2 \left(\frac{q^2}{Mv} \right),$$

where F_1 and F_2 are arbitrary functions. This result has been proved by Bjorken (1969) under certain rather general assumptions. At the present time experiment appears to vindicate this result even for values of $+q^2$ as low as 0.5 GeV^2.

In comparing the data with experiment and bearing in mind the possible relevance of a composite model of the nucleon it is worth examining to

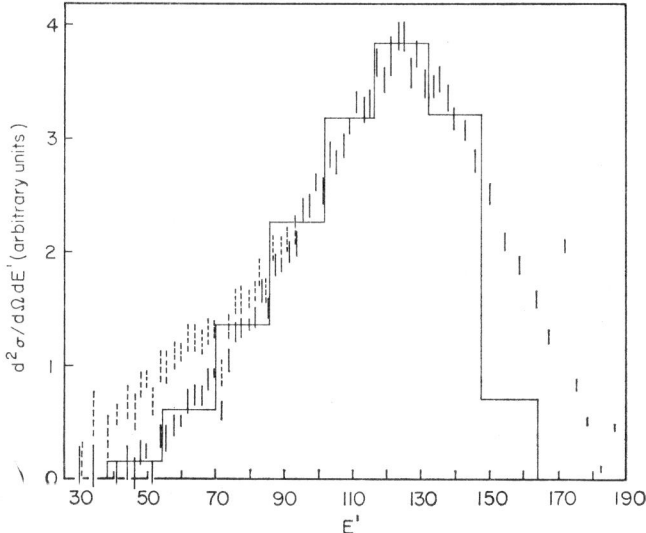

FIGURE 11. The energy loss spectrum $(\partial^2 \sigma)/(\partial \Omega \partial E')$ as a function of final electron energy for 194 MeV electrons scattered inelastically by carbon nuclei through an angle of 135°. The different experimental results | | refer to different estimates of loss of electron energy through radiation. The histogram is the spectrum expected when the proton is supposed to move in a parabolic well potential and to have a gaussian charge distribution. (P. Bounin and G. R. Bishop, 1961.)

begin with the inelastic scattering of electrons by atoms and nuclei. (cf. Gilman, 1969). Examples of these latter are shown in Figures 9, 10 and 11. In Figure 9 the energy loss spectrum of 25 keV electrons on silver atoms is shown for energy losses of the order of 10 eV. These correspond to discrete excitations of the silver atoms and in such cases the electrons undergo small angle scattering only. (The corresponding process for nucleons gives rise to the pronounced peaks for smaller energy losses shown in Figure 12). At large angles and higher energies the contribution from such resonant processes is reduced but in its place one finds scattering by the elementary constituents themselves. In Figure 10 the energy loss spectrum on carbon, copper and gold is shown for 400 keV electrons at 20°. The peaks correspond to collisions in which the incident electron makes an energy momentum conserving collision with an atomic electron (Cooper and Kolbenstvedt, 1972).

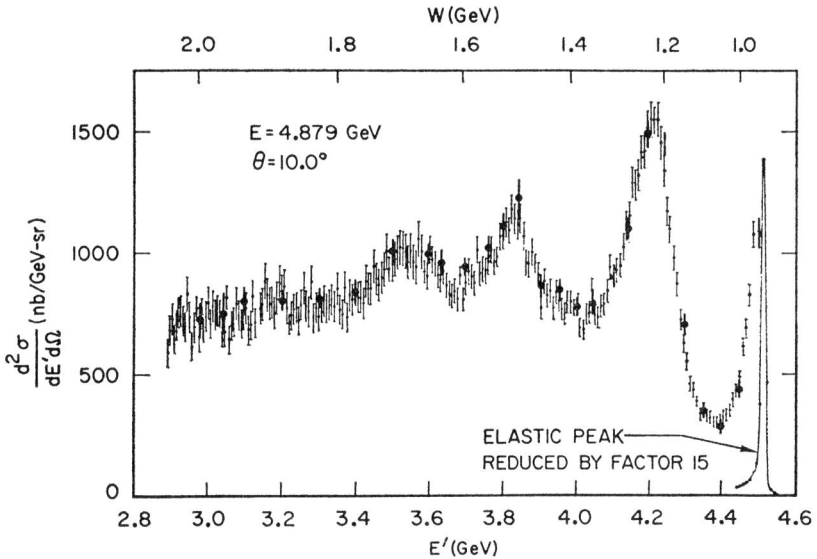

FIGURE 12. The energy loss spectrum $\partial^2\sigma/(\partial\Omega\partial E')$ as a function of final electron energy E' and missing mass W (see text Section VI) for 4.879 GeV electrons scattered inelastically by protons through an angle of 10°. (E. D. Bloom et al., 1970).

In general the final kinetic energy of the inelastically scattered electron making an elastic collision with a target constituent of mass M initially at rest is

$$T' = \frac{(T^2 + 2T)[\cos\theta(M^2 - \sin^2\theta)^{\frac{1}{2}} + \cos^2\theta] - T(1 - M)(1 + M + T)}{(T^2 + 2T)\sin^2\theta + 2TM + (M + 1)^2}$$
(24)

in units in which $m = c = 1$ where m is the electron mass. When $M = 1$, $T' = T_M$, shown in Figure 10. In the case $M \gg 1$, $T \gg 1$

$$T' = \frac{T^2 M \cos\theta + TM(M + T)}{T^2 \sin^2\theta + M(2T + M)}.$$

These must be more or less independent of the nature of the atom so that the peaks all appear in the same place. Note that the peaks are broader for heavier atoms corresponding to the broader momentum distribution of the atomic electrons in these cases. In Figure 11 the energy loss spectrum is shown of 194 MeV electrons on carbon nuclei corresponding to an angle of scatter of 135°; the large peak corresponds to scattering on individual nucleous. The distribution is again rather broad reflecting the momentum distribution of the nuclear protons. Application of the formula quoted above gives $E = 145$ MeV compared with the experimental value of 120 MeV. Of course in this case the assumption that the nucleons are at rest is evidently not a good one.

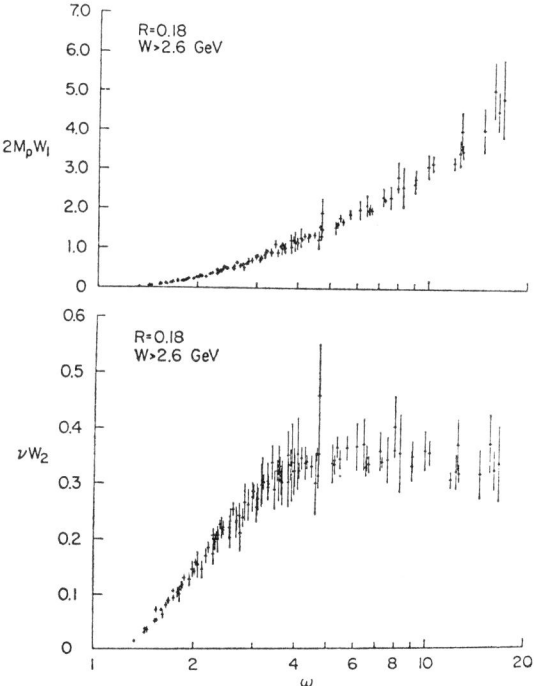

FIGURE 13(a). The function $2M_p W_1$ as a function of $W = 2M_p v/q^2$ for $\sigma_L/\sigma_T = 0.18$ and $W > 2.6$ GeV. (E. D. Bloom *et al.* 1970.) (b) The function vW_2 as a function of $\omega = 2M_p v/q^2$ for $\sigma_L/\sigma_T = 0.18$ and $W > 2.6$ GeV. (E. D. Bloom *et al.*, 1970.)

Turning now to the inelastic scattering on nucleons the resonant transfers already referred to appear at relatively low momentum transfers. At high momentum transfers in view of the uncertainties in the masses and momentum distributions of the hypothetical constituents it is not profitable to apply formula (24) to the energy loss distributions. Instead one uses the data to extract the functions W_1 and W_2 which are shown in Figures 13a and b. The scaling property referred to above is clearly seen. The energy transfers beyond the resonance peaks may then be interpreted as due to scattering on the constituents of the nucleon.

A proposal based on these ideas has been described by Bjorken and Paschos (1969) starting from the "parton" model suggested by Feynman (1969). This views the collision of two strongly interacting particles in the limit of infinite energy in the centre of mass frame as one in which the two oppositely moving particles may radiate primary, independently emitted, quanta termed "partons". Since the centre of mass of the electron and proton at high energy approximates Feynman's conditions, we might hope that such a model of a collection of point-like objects for the proton might have some resemblance to reality. In fact in these circumstances the interaction time is much less than the characteristic times of the partons in the nucleon. Thus the interaction between electron and proton is virtually instantaneous whilst the time dilation effect slows down the parton processes so that in effect they scatter independently. This is reinforced by the requirement of large q^2 which implies that the partons scatter incoherently, that is no phase relation exists between the scattered waves from individual partons.

In this event, the time of interaction τ in the centre of mass frame is of order

$$\tau \sim \frac{1}{q_0} = \frac{4P}{2M\nu - q^2},$$

using centre of mass variables and taking $E_p \sim P$. On the other hand the lifetime of virtual states using uncertainty relation arguments is of order

$$T \sim \frac{1}{((xP)^2 + \mu_1^2)^{\frac{1}{2}} + ((1-x)^2 P^2 + \mu_2^2)^{\frac{1}{2}} - (P^2 + M^2)^{\frac{1}{2}}}.$$

μ_1 and μ_2 are two constituents of the proton of momentum xP and $(1-x)P$, assuming that in the virtual state only 2 partons make up the proton. Then if $p_{1\perp}$ and $p_{2\perp}$ are the transverse components of partons 1 and 2 respectively

$$T \sim \frac{2P}{(\mu_1^2 + p_{1\perp}^2)/x + (\mu_2^2 + p_{2\perp}^2)/(1-x) - M^2)}$$

and so $\tau \ll T$ provided $x \neq 0$ or 1.

ELECTROMAGNETIC STRUCTURE OF NUCLEONS 33

In the approximation $\theta \ll 1$ the cross-section may be written

$$\frac{\partial^2 \sigma}{\partial \Omega \, \partial v} \approx \frac{\alpha^2}{4E^2 \sin^4 \theta/2} W_2(q^2, v) \left[1 + \frac{\sigma_T}{\sigma_T + \sigma_L} \frac{v^2}{2EE'} \right]$$

and since present experimental results are consistent with $\sigma_L/\sigma_T \sim 0$, in the limit of large q^2, we need to consider only W_2.

In the first place for a point particle of unit charge, arbitrary spin and mass M, the function W_2 is given in terms of Dirac's delta function by

$$W_2(q^2, v) = \delta(v - q^2/2M) = M\delta(q.P + \tfrac{1}{2}q^2). \tag{25}$$

In addition we have $W_1 = 0$ for spin 0 and

$$W_1 = \left(1 + \frac{v^2}{q^2}\right) W_2$$

for spin $\tfrac{1}{2}$ with W_1 indeterminate for higher spins. Thus the kinetic energy transfer in the laboratory system for elastic scattering is always $q^2/2M$ in accordance with (25).

We now assume, in following the parton model, that the proton consists of a certain number, N, of free partons with probability $P(N)$ and that the partons move together with the proton with four momentum $p_\mu{}^i = x_i P_\mu$. The parton mass is supposed to remain unchanged. The contribution to W_2 from the ith parton is then $W_2^{(i)}$ where

$$W_2^{(i)} = x_i Q_i^2 \, M\delta(q.x_i P + \tfrac{1}{2}q^2),$$

$$= Q_i^2 \, M\delta\!\left(q.P + \tfrac{1}{2}\frac{q^2}{x_i}\right) = Q_i^2 \, \delta\!\left(v - \frac{q^2}{2Mx_i}\right),$$

where Q_i is the parton charge. One sees immediately that $W_2^{(i)}$ is made up of contributions from point particles of charge Q_i and mass Mx_i. For a general distribution of partons in the proton we have

$$W_2(v, q^2) = \sum_N P(N) \left\langle \sum_i Q_i^2 \right\rangle_N \int_0^1 dx \, f_N(x) \delta\!\left(v - \frac{q^2}{2xM}\right).$$

In this formula $P(N)$ is the probability of finding a configuration of N partons in the proton, $\langle \sum_i Q_i^2 \rangle_N$ is the average of $\sum_i Q_i^2$ for the configuration of N, and $f_N(x)$ is the probability of finding a parton with a fraction x of the proton's momentum in such configurations. Then

$$v\,W_2(v,q^2) = \sum_N P_N \left\langle \sum_i Q_i^2 \right\rangle_N x f_N(x)$$
$$= F(x)$$

where

$$x = \frac{q^2}{2Mv}.$$

It may be shown that the function $F(x)$ satisfies

$$\int_0^1 F(x)\,dx = \sum_N P(N) \left\langle \sum_i Q_i^2 \right\rangle_N \bigg/ N$$

and that $F(x)/x$ is the mean square charge of partons with four momentum $x\,P_\mu$. Experimental results indicate that

$$\int_0^1 F(x)\,dx \approx 0.16.$$

In order to proceed with more detailed models it is necessary to define the function $f_N(x_1 \ldots x_N)$ which is the joint probability of finding partons with fractions $x_1 \ldots x_N$ of the momentum of the proton. In terms of this function

$$f_N(x_1) = \int dx_2 \ldots dx_N f_N(x_1 \ldots x_N) \delta\left(1 - \sum_i x_i\right)$$

with

$$\int_0^1 f_N(x_1)\,dx_1 = 1.$$

The simplest assumption concerning $f_N(x_1 \ldots x_N)$ is

$$f_N(x_1 \ldots x_N) = \text{constant},$$

so that

$$f_N(x) = (N-1)(1-x)^{N-2}.$$

To introduce an element of (probably quasi!) reality, Bjorken and Paschos treat the cases of a proton consisting of 3 quarks, and of 3 quarks together with a background of quark antiquark pairs (mesons). (For a brief account of quarks and their place in the system of elementary particles see Eden, 1970). The results are as follows

(a) 3 quarks
$$vW_2 = F_3(x) = 2x(1-x).$$

(b) 3 quarks plus background
$$F_p(x) = \frac{1}{1-\ln 2}\left\{\frac{2(1-x)}{9(2-x)} + \frac{1}{6}\frac{x}{(1-x)^2}\left[\ln\frac{2-x}{x} - 2(2-x)\right]\right\},$$

for the proton and
$$F_n(x) = \frac{2}{9}\frac{1}{1-\ln 2}\left(\frac{1-x}{2-x}\right).$$

for the neutron.

To obtain the last result it is assumed that

(i)
$$\frac{\langle \sum_i Q_i^2 \rangle_{\text{backgr}}}{N} = \tfrac{1}{3}\{(\tfrac{2}{3})^2 + (\tfrac{1}{3})^2 + (\tfrac{1}{3})^2\} = \tfrac{2}{9},$$

using the usual quark charges with equal probabilities and

(ii)
$$P(N) = \frac{C}{N(N-1)}.$$

This last has the asymptotic dependence on N required to give

$$vW_2 \to \text{const} \quad \text{as} \quad x \to 0 \quad (v \to \infty, q^2 \text{ fixed}).$$

These results give

$$\int_0^1 F(x)\,dx = \tfrac{2}{9} + \tfrac{1}{3}\langle 1/N\rangle, > 0.22 \quad \text{for the proton}$$

and

$$= \tfrac{2}{9} = 0.22 \quad \text{for the neutron.}$$

The corresponding plot of vW_2 which is given in Figure 14 should be compared with the experimental data in Figure 13b.

Obviously the theoretical result is strongly model dependent as remarked by its authors. The latest experimental results do however appear at least to indicate a maximum in vW_2. With regard to the parton spin, the fact that $R \equiv \sigma_L/\sigma_T \sim 0$ referred to above, implies that the partons have spin $\tfrac{1}{2}$.

This last result, viz. that $R \sim 0$, also has a bearing on the proton compositeness condition. West (1971) has shown on the basis of rather general conditions, together with the assumption that if $q^2 \sigma_L \to 0$ as

$q^2 \to +\infty$, then $Z_2 = 0$ (see the discussion in Section V). The experimental smallness of R is thus a further piece of evidence in favour of the composite proton.

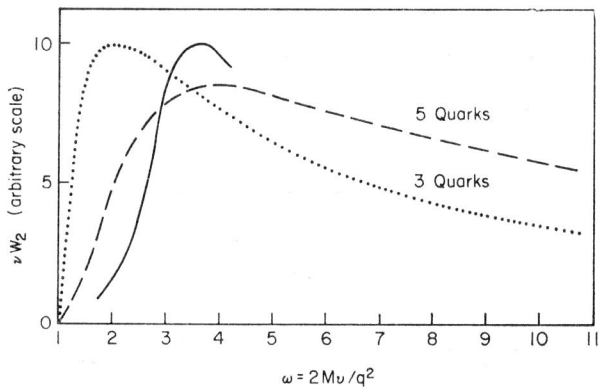

FIGURE 14. Theoretical estimates of νW_2 as a function of $\omega = 2M_p\nu/q^2$ quoted by E. D. Bloom et al. (1971).
⋯⋯ J. D. Bjorken and E. A. Paschos (1969).
—— J. D. Bjorken (1968).

VII. High energy pp elastic scattering and proton structure

It is possible to pursue the particulate model of the proton further by considering the elastic scattering of protons by protons. Van Hove (1967) has discussed this by treating the protons as an assembly of point-like quarks although for the present purpose any particulate structure would give the same result.

The result for the scattering amplitude is of the form

$$f(t) = ig(t)\, G^2(t). \tag{26}$$

Here $g(t)$ is the quark-quark scattering amplitude and $G(t)$ the electric form factor. For small values of t, $g(t) \sim g(0)$ since the quarks are assumed small in size compared with the size of the proton. The quantity $G(t)$ appears in this model as an overlap effect of the quark wave functions of the two nucleons. When this is summed up over the two nucleons assuming no multiple scattering effects, the result (26) is obtained. The agreement with experiment is rather good for small t.

Wu and Yang (1965) and Chou and Yang (1968) have found a similar result starting from a droplet model of the nucleon and treating the collision as one in which two spherically symmetric opaque matter distributions pass through one another. To each sphere the other appears effectively as a disc with opacity given by

$$D(x, y) = \int_{-\infty}^{\infty} \rho(x, y, z)\, dz = D(\mathbf{b}),$$

where ρ is taken to be the charge distribution. In a collision the opacity at impact parameter b is given by

$$F(\mathbf{b}) = \int D(\mathbf{b} - \mathbf{b}') D(\mathbf{b}')\, d^2\mathbf{b}'. \tag{27}$$

The transmission factor, at impact parameter $\mathbf{b}$, is

$$S(\mathbf{b}) = \exp\left[-\text{const}\, F(\mathbf{b})\right] \tag{28}$$

and the differential scattering cross-section for scattering through the angle θ is

$$\frac{d\sigma}{dt} = \pi |f(s, t)|^2,$$

where

$$f(s, t) = i \int_0^{\infty} (1 - S(b))\, J_0(b\sqrt{-t})\, b\, db$$

and

$$s = -(p_1 + p_2)^2.$$

p_1 and p_2 being the four momenta of the incoming protons,

$$t = -2k^2(1 - \cos\theta)$$

and k is the momentum of either proton in the centre of mass system.

It may be shown that if

$$G(t) = \frac{1}{2\pi} \int D(\mathbf{b}) \exp(i\mathbf{k} \cdot \mathbf{b})\, d^2\mathbf{b},$$

where $\mathbf{k}$ is a vector in the impact parameter plane with magnitude $\sqrt{-t}$ then

$$G^2(t) = 2\pi \int_0^{\infty} F(b)\, J_0(\sqrt{-t}\, b)\, b\, db.$$

Furthermore for large $|t|$

$$f(s, t) \to \text{const}\, G^2(t).$$

Thus for large s and t, the pp differential scattering cross-section should be proportional to the fourth power of the proton charge structure factor.

In fact, bearing in mind the discussion given by Van Hove, we should expect multiple scattering effects to be important at large t and indeed Chou and Yang have formulated their model as a multiple scattering expansion. They do indeed find that multiple scattering effects are significant for large t and are just what is required to bring theory and experiment into agreement. The connection between the liquid drop and particulate models has been considered by Byers and Frautschi (1970) using the Glauber Multiple scattering approximation. They find that the main approximation made in the Chou–Yang model consists in ignoring correlations between the particle like constituents of the proton. Indeed this can be plainly seen from (27) and (28). The Chou Yang procedure consists essentially of a power series expansion of the exponential in (28).

Comparison between the form factor deduced in this way and that found experimentally in electron proton scattering is given in Figure 6b.

VIII. Conclusions

The use of electrons as a probe in determining the electromagnetic structure of nucleons has provided a vast amount of data concerning what one might call the static structure of the neutron and proton, that is the charge and magnetic moment distributions. Loosely speaking this is now well documented down to distances of the order 10^{-14} cm. In a particular inertial frame of reference (the Breit frame which depends on the momentum transfer, see Section III) the two distributions are remarkably similar for the proton and they have the form deduced from the dipole formula

$$\rho(r) = a\,e^{-\alpha r},$$

where $\alpha = 4.23\,\mathrm{f}^{-1}$ for $r \gtrsim 10^{-14}$ cm.

The neutron magnetic moment distribution differs from this only in that the neutron distribution is μ_n/μ_p times the proton magnetic moment distribution to a good approximation.

Another way of representing the elastic data is to introduce the mean squared radius of the charge distribution through the formula

$$G_E(q^2) = 1 - \tfrac{1}{6}\langle r^2\rangle q^2 + \ldots.$$

The value of $\langle r^2\rangle^{\frac{1}{2}}$ from the slope of the graph for $G_E(q^2)$ is 0.813 f. From the dipole formula the corresponding value is 0.822 f. The reasons for this value and for the fact that a remarkably similar result can be obtained from an analysis of proton–proton scattering which involves strong interactions are not known. The fact that the distribution falls off faster than $1/q^2$ does however imply that the simple meson cloud model is not true. One model which requires a faster fall off is the composite nuclear model, but again nothing is known of the nature of the constituents.

Turning now to the inelastic scattering or the dynamic structure of the nucleon, two features are immediately apparent from the data. The first is the existence of quasi discrete excited states of the nucleon giving rise to peaks in the electron energy loss spectrum exactly as shown by electrons in gases or solids. The second is the existence of what might be called quasi elastic scattering i.e. a substantial tail in the energy loss distribution very similar to that found in electron scattering off atoms and nuclei and attributed there to scattering off the atomic or nuclear constituents behaving as free independent particles. All this again seems highly suggestive of a composite model and leaves open the nature of the constituents. For the moment they are simply designated as partons which leaves their identity wide open. Are the partons quarks? Can they exist independently of nucleons or are they some manifestation of a collective nucleonic excitation? These and many other questions are for the future. All one can say at present is that we seem tantalisingly close to major advances in our knowledge of the ultimate constitution of matter. Perhaps the data from the higher energy accelerators in the not too distant future will provide some of the answers.

Acknowledgements

Thanks are due to the Stanford Linear Accelerator Centre for a copy of their Annual Report for 1970 and reprints of their work on High Energy Inelastic Scattering.

Bibliography

Drell, S. D. and Zachariasen, (1961). Electromagnetic Structure of the Nucleon, O.U.P. Oxford.

Gourdin, M. (1966). Diffusion des Electrons de Hautes Energies, Manon et Cie, Paris.

References

Ball, J. S. and Zachariasen, F. (1968). *Phys. Rev.* **170**, 1541.
Bartel, W., Dudelzak, B., Krebbiel, H., McElroy, J., Meyer-Berkhaut, U., Morrison, R. J., Nguyen-Ngoe, H., Schmidt, W. and Weber, G. (1967). *Phys. Lett.* **25B**, 236.
Bartel, W., Biisyer, F. W., Dix, W. R., Felst, R., Harris, D., Krebsbiel, H., Kuhlman, P. E., McElroy, J. and Weber, G. (1970). *Phys. Lett.* **33B**, 245.

Berger, Ch., Burkert, V., Knop, G., Lanjenback, B. and Rith, K. (1971). *Phys. Lett.* **35B**, 87.
Bjorken, J. D. (1968). *In* 'Selected Topics in Particle Physics', Proc. Intern. School of Physics 'Enrico Fermi', Course XLI Ed. by J. Steinberger, Academic Press, Inc. New York.
Bjorken, J. D. (1969). *Phys. Rev.* **179**, 1547.
Bjorken, J. D. and Paschos, E. A. (1969). *Phys. Rev.* **185**, 1975.
Bloom, E. D., Buschhorn, G., Coterell, R. L., Coward, D. H., De Staebler, H., Drees, J., Jordan, C. L., Miller, G., Mo. L., Piel, H., Taylor, R. E., Breidenbach, M., Ditzler, W. R., Friedman, J. I., Hartmann, G. C., Kendall, H. W. and Pouchet, J. S. (1970). *Proc. of XV International Conf. on High Energy Physics,* Kiev, U.S.S.R.
Boersch, H., Geiger, J., Hellwig, H. and Michel, H. (1962). *Z. Phys.* **169**, 252.
Bounin, P. and Bishop, G. R. (1961). *J. Phys. Radium,* **22**, 555.
Budnitz, R. J., Appel, J., Carroll, L., Chen, J., Dunning, J. R. Jr., Gostein, M., Hanson, K., Imrie, D., Mistrette, C., Walker, J. K. and Wilson, R. (1968). *Phys. Rev.* **173**, 1357.
Byers, N. and Frautschi, S. (1970). *In* 'Quanta' P. G. O. Freund, C. J. Goebel and Y. Nambu (eds). University of Chicago Press, Chicago and London.
Cassiday, G. L. (1971). *Phys. Rev.* **D3**, 1109.
Chou, T. T. and Yang, C. N. (1968). *Phys. Rev.* **170**, 1591.
Cooper, J. W. and Kolbenstvedt, H. (1972). *Phys. Rev.* **A5**, 677.
Cornwall, J. N. and Levy, D. J. (1969). *Phys. Rev.* **178**, 2356.
Coward, D. H., Destaebler, H., Early, R. A., Litt, J., Minten, A., Mo, L. W., Panofsky, W. K. H., Taylor, R. E., Breidenbach, M., Friedman, J. I., Kendall, H. W., Kirk, P. N., Barish, B. C., Mar, J. and Pine, J. (1968). *Phys. Rev. Lett.* **20**, 292.
Eden, R. J. (1970). *In* 'Essays in Physics', (G. K. T. Conn and G. N. Fowler, eds), Vol. 1, p.1. Academic Press Inc. (London) Ltd., London and New York.
Feynman, R. P. (1969). *Phys. Rev. Lett.* **23**, 1415.
Fowler, G. N. and Wolfendale, A. W. (1957). *Nuc. Phys.* **3**, 299.
Frazer, W. R. and Fulco, J. R. (1960). *Phys. Rev.* **117**, 609.
George, E. P. and Evans, J. (1950). *Proc. Phys. Soc.* **A63**, 1248.
Gilman, F. J. (1969). Fourth International Symposium on Electron and Photon Interactions at High Energies, 1969, Daresbury Nuclear Physics Lab., U.K.
Gourdin, M. (1963). *Nuovo Cim.* **28**, 533, (1964). **32**, 493 and (1965). **35**, 1105.
Hand, L. N., Miller, D. G. and Wilson, R. (1963). *Rev. Mod. Phys.* **35**, 335.
Kroll, N. K., Lee, T. D. and Zumino, B. (1967). *Phys. Rev.* **157**, 1376.
Litt, J., Buschhorn, G., Coward, D. H., Destaebler, H., Mo. L. W., Taylor, R. E., Barish, B. C., Loken, S. C., Pine, J., Friedman, J. I., Hartmann, G. C. and Kendall, H. W. (1970). *Phys. Lett.* **31B**, 40.
Missoni, G., Dick, C. E., Placious, R. C. and Moty, J. W. (1970). *Phys. Rev.* **A2**, 2309.
Neal, R. B. (1968). Ch. (5). The Stanford Two-Mile Accelerator, W. A. Benjamin Inc., New York, N.Y.
Pipkin, F. M. (1970). *In* 'Essays in Physics', (G. K. T. Conn and G. N. Fowler, eds), Vol. 2., p.1. Academic Press Inc. (London) Ltd., London and New York.
Rutherglen, J. G. (1969) Fourth International Symposium on Electron and Photon Interactions at High Energies, 1969, Daresbury Nuclear Physics Lab., U.K.
Sachs, R. G. (1962). *Phys. Rev.* **126**, 2256.

Sakurai, J. J. (1960). *Ann. Phys.,* **11,** 1.
Van Hove, L. (1967). Particle Interactions at High Energies (T. W. Preist, and L. L. J. Vick, eds). Ch. 2, Oliver and Boyd, Edinburgh and London.
West, G. B. (1971). *Phys. Rev. Lett.* **27,** 762.
Wilson, R. (1967). Particle Interactions at High Energies (T. W. Priest, and L. L. J. Vick, eds), Ch. 3, Oliver and Boyd, Edinburgh and London.
Wilson, R. R. and Levinger, J. S. (1964). *A. Rev. Nucl. Sci.* **14,** 135.
Wu, T. T. and Yang, C. N. (1965). *Phys. Rev.* **137B,** 708.

Phase Transitions, Symmetry and Dimensionality†

MICHAEL E. FISHER

Baker Laboratory, Cornell University, Ithaca, New York, U.S.A.

I.	Introduction	43
II.	Phase diagrams and order	44
III.	Long-range and spontaneous order in magnets	47
IV.	Symmetries	53
V.	The sharpness of a transition	55
VI.	Existence and non-existence of a transition	59
VII.	Transitions in one dimension	62
VIII.	Dimensionality and critical behaviour	69
IX.	Absence of ordering in two-dimensional isotropic systems	70
X.	Two dimensions in the real world?	77
XI.	Postscript: 1972	86
Acknowledgements		88
Bibliography		88
References		88

I. Introduction

My theme in this survey is the influence of symmetry on phase transitions and, in particular, the influence of dimensionality. The discussion will be somewhat of a whirlwind tour in which one visits all the main capitals of Europe in the space of a week but, hopefully, sees sights and gathers impressions which tempt one to go back for a longer and more relaxed visit.

I will begin by outlining some of the general questions; first there is the matter of the *shape* of the phase diagram. This, of course, is one of the central problems when one is thinking about phase transitions. Then I will characterize the different sorts of *order* one can talk about. These concepts

---

† Based on a lecture presented at the Chicago Solid State Colloquium Series on 13th April, 1970.

of order are the tools of the theoretician, and increasingly of the experimentalist, in thinking about phase transitions. I will then stress the fact that phase transitions, when you do good experiments, are very sharp. From a theoretical point of view this means that one must consider the so-called *thermodynamic limit*; discussing that limit carefully is not an entirely idle problem, as I will point out. We will then examine what one knows about the origins of phase transitions, emphasizing the point that if one wants some sort of phase transition—if one wants a gas to condense—there must be attractive forces, otherwise the atoms will not particularly like being next to one another! But, despite the fact that various distinguished physicists on occasion have thought that it was enough to have attractions in order to have phase transitions, that is only *half* the story. In a number of cases, considerations of dimensionality or of symmetry prevent a true transition even though there are attractions. In this respect there has been appreciable progress during the last few years. I will finish by concentrating attention on critical phenomena and what happens near a critical point. However, I will emphasize solids and underplay liquids. In particular, I want to discuss a magnetic critical point, where recently one has been able to look, in the real world, at apparently two-dimensional or rather closely two-dimensional systems. Some quite beautiful experiments have been performed in the last year or two, and I would like to review these briefly.

II. Phase Diagrams and Order

Let us start then with the question of the phase diagram. For a single component system, this is a graph of pressure p versus temperature T. (See Figure 1). In the diagram there are various lines, or phase boundaries, separating the different phases. At low pressures and temperatures we always have a gas; at low temperatures, of course, the gaseous phase generally condenses directly into a solid or crystalline phase when p increases. Then there is a *melting line* from solid to liquid and a third line ending in a *critical point* (marked by an open circle). For simplicity we ignore the possibility of many solid phases. Above this third line, the *vapor pressure line*, we call the material "liquid", but all the evidence is that there is really no distinction between a liquid and a gas. I can go over the top of the critical point from one to another (see the double arrow in Figure 1) without anything untoward happening; everything remains completely smooth and analytic. That happily disposes of the problem of liquids—they are just dense gases! They may be a little more awkward to compute, if we insist on learning the last decimal place; but from the point of view of phase transitions, they are not especially interesting.

The same is not true of the liquid–solid transition line or melting line, and one of the first questions is a very old and famous one: "If we can get up to a high enough pressure, should we expect to find some critical point on the melting line?" The answer to that is: "No, you should not." The reason why goes to the heart of the way one thinks about phase transitions. One follows the same line of thought as when asking what is the difference between liquid and gas, which was, "None in principle". In both cases the arrangement of the atoms is essentially random on a large scale or uncorrelated at large distances. On the other hand, in a solid we know that the atoms are arranged regularly. There may be defects every now and then, but nevertheless, out to macroscopic distances we have a regular arrangement. The solid can thus be characterized by looking at the net correlation function,

$$G(\mathbf{r}; p, T) = g_2(\mathbf{r}) - 1, \qquad (1)$$

where $g_2(\mathbf{r})$ is the standard pair correlation function which is a measure of the probability that if I put one atom at the origin, I will find another one at a distance $\mathbf{r}$. In a fluid $G(\mathbf{r})$ falls quite rapidly to zero as r increases. By contrast, in a crystal one expects this function to oscillate about zero out to infinity—even at infinity one expects to find a finite amplitude of oscillation. So our criteria for having a crystalline phase is that the net correlation function $G(\mathbf{r})$

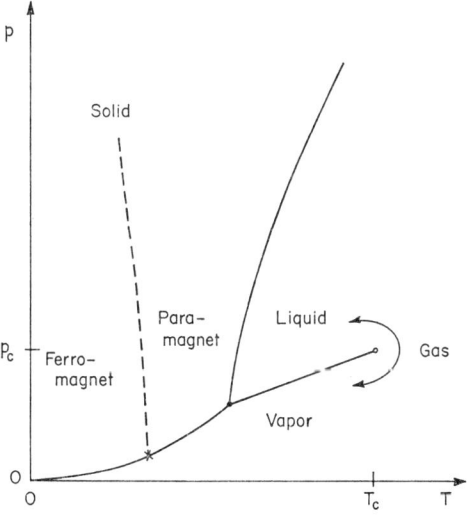

FIGURE 1. Phase diagram of a simple single-component system. (The open circle denotes the critical point. The cross indicates the Curie point for ferromagnetism.)

does not tend toward zero as r tends to infinity. In other words, there is *long-range order*. One could, for example, take the limiting amplitude of the oscillation far from the origin as a measure of that long-range order. So what characterizes a solid is its long-range order. Why does that mean that we cannot have a solid–liquid critical point? Well, suppose that there were a solid–liquid critical point: then, as in the case of the gas–liquid critical point, one could find a path, not crossing any phase boundary, from liquid with no long-range order to solid with an infinitely oscillating $G(\mathbf{r})$. Since there is a definite qualitative distinction between liquid and solid—the presence of long-range order—any such path must cross *some* locus separating liquid from solid. This locus can be no more than the continuation of the melting line, which hence cannot come to an end; (although, it could of course curve back and meet the p-axis at $T = 0$ or intersect some other phase transition line in a triple point). The melting line might conceivably change its character at some point above which the melting transition, instead of being of a first-order character with a density discontinuity, becomes a continuous transition. Essentially, there has to be some line—the melting curve cannot end at a critical point.

I believe these arguments to be, in essence, incontrovertible. The only way to refute them is to say that you do not like my definition of a crystal. In that there may be something real to argue about. Indeed, I have already sold you a piece of folklore which maybe you should put down as a bad bargain. Habitually, one talks about the pair correlation function $g_2(\mathbf{r})$, which is what can be found from scattering experiments. In point of fact, however, I am not convinced that knowing $g_2(\mathbf{r})$ is enough to define unambiguously a particular crystal. To perform a scattering experiment one normally takes a crystal and holds it still; the diffraction pattern with its characteristic spots is then essentially the Fourier transform of $G(\mathbf{r})$. But when you think further about this and about the statistical mechanical definition of $g_2(\mathbf{r})$ and allow for the fact that noncentrosymmetric crystals are known, doubts grow. My tentative conclusion has been that to be sure you have a crystal and know what sort it is, you need to look at the five-particle correlation-function $g_5(\mathbf{r}_{21}, \mathbf{r}_{31}, \mathbf{r}_{41}, \mathbf{r}_{51})$. Thus, I need four points to hold the crystal fixed the right way up, and I need one point to look far away to make sure it has not melted at large distances from the origin i.e. that there really is long-range order. Maybe someone can prove that I can be certain of crystalline structure with a correlation function for a smaller number of particles, but I am doubtful. The reason we do not talk about this is that g_5 is not a very handy thing to measure! The fact that this is not a trivial problem becomes clear when one extends the phase diagram of Figure 1 to include an extra phase present in many systems with sufficiently complex molecules. I am referring to *liquid crystalline phases* which are currently of appreciable technological interest and under intensive study. The liquid crystal phase occurs between solid and liquid, as illustrated in Figure 2. Again it is an "enclosed" phase i.e. there is no critical point either to solid or to normal liquid. A liquid crystal is certainly characterized by some sort of long-range order but this is *not* complete crystalline order. Basically, liquid crystals have directional order (their molecules always have an axis) but they have relatively little positional order. However, I believe that it is not yet clear precisely what sorts of order do characterize the different types of liquid crystal. Nevertheless, we again expect the existence of each phase to be associated with a definite and unique type of order.

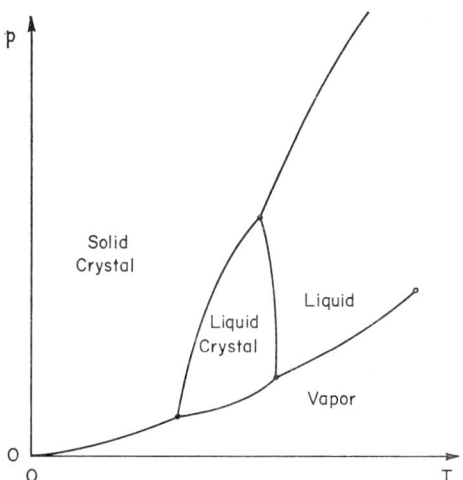

FIGURE 2. Phase diagram illustrating location of aliquid crystalline phase (schematic).

III. Long-range and Spontaneous Order in Magnets

To develop these concepts of order in more detail and to expose some of the different theoretical criteria, I shall focus attention on magnetic systems which are conceptually simpler to think about. Systems which have magnetic phase transitions are essentially all solids: as indicated in Figure 1, a crystal at low temperatures may be ferromagnetic. If, as usual, the crystal is observed under its vapor pressure, the ferromagnetic state ends at the Curie point (shown by a cross in Figure 1) above which the crystal is paramagnetic. Under pressure the Curie point is drawn out into a Curie-line (shown dashed in Figure 1) to the low temperature side of which the compressed crystal remains ferromagnetic with a spontaneous magnetization. Figure 3 illustrates the temperature variation of the spontaneous magnetization. Actually, I am cheating somewhat since what is shown really represents the spontaneous *sub-lattice* magnetization curve of an antiferromagnet, namely manganese fluoride—very famous data obtained by Heller and Benedek (1962). What they actually measured was a resonance frequency (curve with data points) which is proportional to the spontaneous sub-lattice magnetization. It vanishes very abruptly, in this case, at a fairly low temperature. The figure contrasts the experimental data with the molecular field theory prediction, which we note in passing. I want to characterize the order present and distinguish the various features to be considered. Let us think of a magnet; antiferromagnet and ferromagnet will not be too different. Of the variety of quantities one may discuss, the easiest to think about is the spontaneous magnetization or, if I have an antiferromagnet, the magnetization of a sub-lattice. To define that precisely, consider the magnetization $M(H, T)$ with a field H present at some temperature T. Now reduce the field

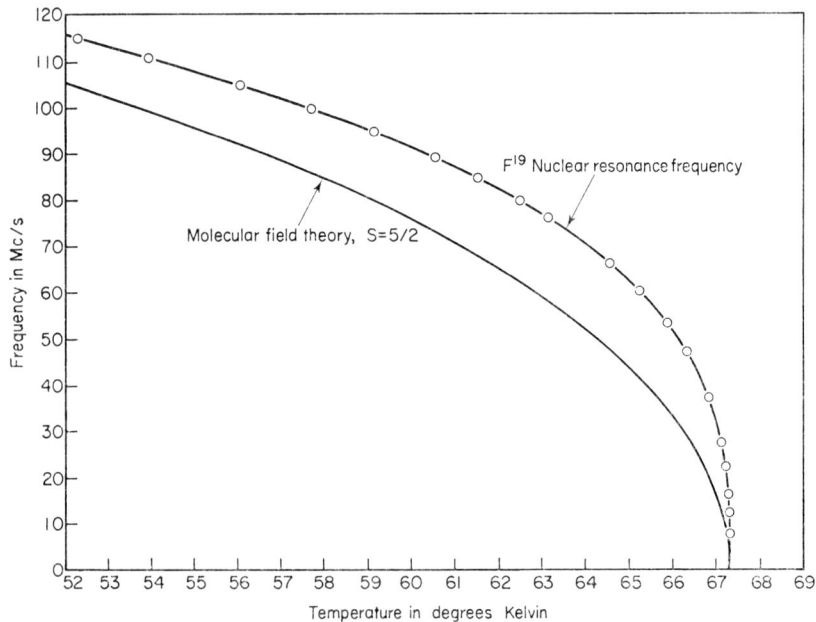

FIGURE 3. The sub-lattice magnetization of MnF_2 as measured by nuclear magnetic resonance (Heller and Benedek, 1962).

to zero. If we are below the ordering temperature, we will obtain the spontaneous magnetization, namely,

$$M_0(T) = \lim_{H \to 0+} M(H, T). \qquad (2)$$

The M versus H plot is discontinuous at $H = 0$; if I took the limit from $H < 0$, I would obtain minus the previous result. For an antiferromagnet one simply considers $M'(H', T)$ the magnetization of one sub-lattice in a *staggered* magnetic field H', which is positively directed on the one sub-lattice but negatively on the other.

The expression (2) represents as it were, the classic description of the ordering that is going on. For a crystal one can also give a definition of this form; it is a bit more tricky to write a precisely similar expression—the crystal is more closely analogous to an antiferromagnet—and I will not take the time to do it. Before leaving manganese fluoride, however, let me just point out what is found close to T_c where $M_0'(T)$ vanishes: the behavior is accurately described by

$$M_0'(T) \approx A|T - T_c|^\beta, \qquad T \to T_c-, \qquad (3)$$

where the characteristic exponent β is about 0.33 for real, three-dimensional specimens. This can be seen from Figure 4 where the cube of the resonance

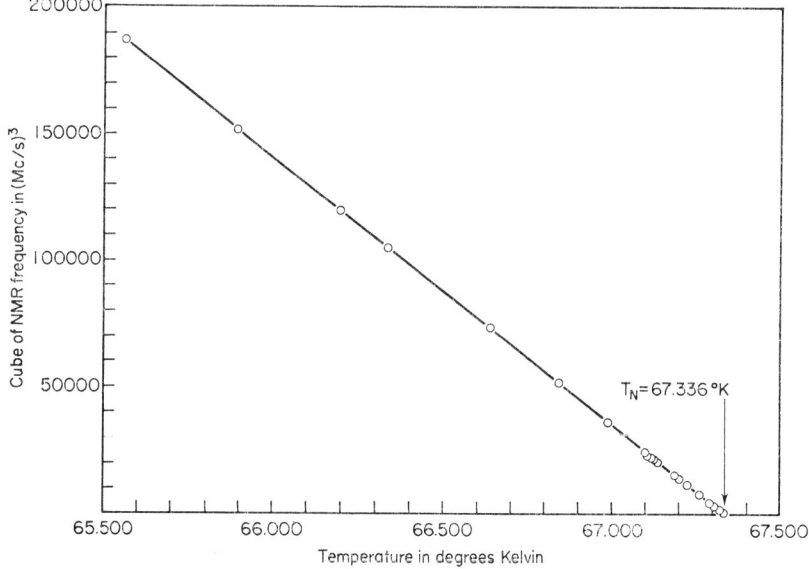

FIGURE 4. Cube of the sub-lattice magnetization of MnF_2 versus T demonstrating that $\beta \simeq 0.33$ (Heller and Benedek, 1962).

frequency appears linear with respect to T. Mean field theory yields the incorrect prediction $\beta = \frac{1}{2}$.

One of the theoretical difficulties is actually proving things about the magnetization $M(H, T)$, or staggered magnetization $M'(H', T)$. For simplicity, consider a system of localized spins $\mathbf{S_r}$ on lattice sites $\mathbf{r}$. We may restrict attention to magnetization in, say, the z direction and consider one typical spin, say at the origin. Then we have

$$M(H, T) = \langle S_0{}^z \rangle, \tag{4}$$

where the expectation is taken in field H at temperature T. This may be a convenient quantity to examine but is not always so, particularly in an antiferromagnet. Moreover, in considering the theory of scattering experiments one is led naturally to consider another measure of the order, namely, the mean square order $M_\sigma{}^2$. In addition, the theory of diffuse scattering leads one to study what is sometimes called the *short-range order*, or the actual spin correlations themselves. Consider first these correlations which are the analogues of those introduced in discussing fluids and crystals. The correlation of the z-components of a spin at site $\mathbf{r}$ with one at site $\mathbf{r'}$ is

$$\sigma(\mathbf{r}, \mathbf{r'}; T) = \langle S_\mathbf{r}{}^z S_{\mathbf{r'}}{}^z \rangle, \tag{5}$$

where we suppose the field H is zero. If the limit

$$\sigma(\infty; T) = \lim_{|\mathbf{r}-\mathbf{r'}| \to \infty} \sigma(\mathbf{r}, \mathbf{r'}), \tag{6}$$

exists and is nonzero, we have long-range order; as a matter of fact, this is really something we should call the *short* long-range order—I want to give you some flavor of the theoretical distinctions here. There is indeed another quantity called *long* long-range order, and it turns out that neither of these is that easy to discuss theoretically; and they can be different.

The short long-range order really is tied to what one sees when one does a Bragg scattering experiment. If there is short long-range order—that is if the spin pair-correlations do not decay to zero but decay to some nonzero $\sigma(\infty)$—then that implies essentially a sharp Bragg peak. The peak will be at the origin and its images in reciprocal space if (6) applies; but if there is some oscillation so that $\sigma(\mathbf{r}, \mathbf{r'}) \approx \sigma(\infty) \cos [\mathbf{q} \cdot (\mathbf{r} - \mathbf{r'})]$ as $|\mathbf{r} - \mathbf{r'}| \to \infty$, then the Bragg peak will be at a momentum-transfer $\mathbf{q}$. In both cases the intensity of the peak is proportional to the short long-range order parameter $\sigma(\infty; T)$.

There is a "folklore theorem" which states that $M_0(T)$ is related to $\sigma(\infty; T)$. We can see that there should, perhaps, be such a relation, by considering the other form of order mentioned, namely M_σ, the square-root of the mean-square order. This is defined by

$$M_\sigma^2 = \left\langle \left(N^{-1} \sum_{\mathbf{r}} S_{\mathbf{r}}^z\right)^2 \right\rangle, \qquad (7)$$

where the sum runs over all the N spins in the system. It may also be fruitful to consider the order in a subdomain by restricting the sum to an appropriate subset of all the spins.) To indicate how the argument goes—although I must also point out that it is *not* rigorous—let us multiply out the squared sum in (7) to obtain

$$\begin{aligned} M_\sigma^2 &= N^{-2} \sum_{\mathbf{r}} \sum_{\mathbf{r}'} \langle S_{\mathbf{r}}^z S_{\mathbf{r}'}^z \rangle, \\ &= N^{-2} \sum_{\mathbf{r}} \sum_{\mathbf{r}'} \sigma(\mathbf{r}, \mathbf{r}'). \end{aligned} \qquad (8)$$

If we have a large system we can expect the correlation function $\sigma(\mathbf{r}, \mathbf{r}')$ to depend only on the difference of the arguments, that is,

$$\sigma(\mathbf{r}, \mathbf{r}') \simeq \sigma(\mathbf{r} - \mathbf{r}'). \qquad (9)$$

At best this is an approximation valid throughout the bulk of the system; it must fail near the edges or boundaries. At the cost of this approximation we perform one of the summations in (8) which thereby gives

$$M_\sigma^2 \simeq N^{-1} \sum_{\mathbf{r}} \sigma(\mathbf{r}). \qquad (10)$$

Now we argue as follows: If $\sigma(\mathbf{r})$ decays not to zero but to some $\sigma(\infty)$, then for a large system the sum in (10) will be dominated by what happens at large $\mathbf{r}$. (The factor N^{-1} "damps" out the significance of the short range behaviour.) Thus we have

$$M_\sigma^2 = \sigma(\infty) + e_N, \qquad (11)$$

where the error

$$e_N \simeq N^{-1} \sum_{\mathbf{r}} [\sigma(\mathbf{r}) - \sigma(\infty)] \qquad (12)$$

is, optimistically, small and vanishes as $N \to \infty$. In the thermodynamic limit we hence conclude

$$M_\sigma^2 = \sigma(\infty), \quad [?] \qquad (13)$$

which relates the mean-squared order to the long-range order. However, I repeat that this statement is not easy to prove properly (as indicated by the query). (We may also note that what is really entering in (11) and (12) is the *long* long-range order since the thermodynamic limit is effectively being taken *after* the long-range limit.)

Lastly, to relate M_σ to the spontaneous magnetization M_0 we appeal to another "folk theorem" which says that in the ordered state the whole ferromagnetic crystal points either "up" or "down". This is really a fallacy because most magnetic systems break up into domains, half pointing up and half pointing down. If we overlook that little detail, the argument runs as follows: the probability distribution of the total magnetization when $T < T_c$ will be *double-peaked*, symmetric about $M = 0$ with peaks located at $M = \pm M_0$ (see the solid line in Figure 5). Since these peaks are supposed to be very tall and very narrow we expect

$$M_0 = M_\sigma = \sqrt{[\sigma(\infty)]}, \quad [?] \tag{14}$$

where the last equality just restates (13). Again this expected equality is not necessarily valid. I have already implied one difficulty with the argument, namely, the distribution of magnetization need not be double-peaked on a "thermodynamic scale" since if I merely insert one domain wall into the sample, (which costs only a "surface free energy"), I can obtain any total magnetization intermediate between $-M_0$ and $+M_0$. The area between the peaks is thus effectively filled in, see the dashed line in Figure 5.

My aim in indicating some of these arguments is to bring out various theoretical criteria for ordering and to point out that they can be quite elaborate. The elaboration is necessary if one really wants to prove anything; later, I shall quote some rigorous theorems about ordering.

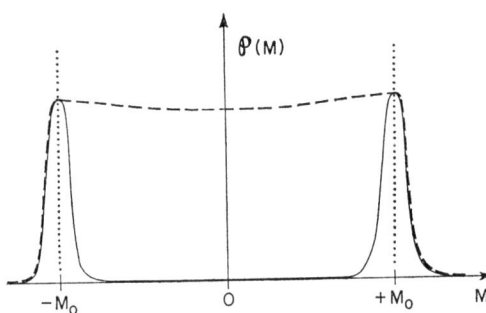

FIGURE 5. Schematic plot of the distribution of the total magnetization in an ordered ferromagnet. The solid line denotes the customary "double peak" expectation; the dashed line is a more realistic guess.

IV. Symmetries

Now, let me consider another aspect of phase transitions, namely *symmetry*. In a rather trivial way, in a ferromagnet, you can always argue that there should be no spontaneous magnetization since the system has no reason to prefer the "up" to the "down" direction. That was, in fact, why, in the definition (2) of $M_0(T)$, I had to bring in the limit H tending to zero which "breaks the symmetry". So, this is a trivial up-down symmetry which has been broken. We may take the spontaneous magnetization as a measure of the order parameter, i.e. as a measure of the degree to which we have broken the symmetry. Above T_c, of course, the full symmetry argument applies. It says that we should have a symmetric M versus H curve which, by continuity, must pass through the origin so that $M = 0$ when $H = 0$.

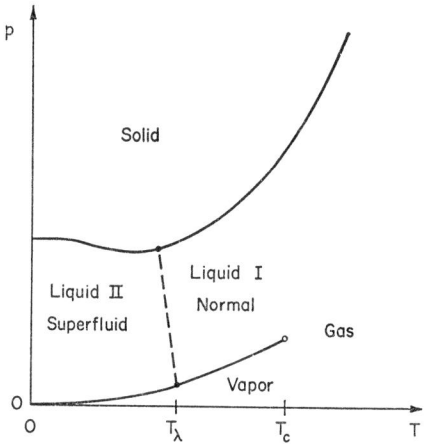

FIGURE 6. Schematic phase diagram for helium showing the enclosed superfluid phase.

There are more significant aspects to the symmetry of an order parameter, which we see as soon as we go to more complicated phase diagrams. Let me just remind you of a phase diagram of particular interest: that phase diagram in which there appears an additional liquid phase at low temperatures, namely liquid He II, a superfluid phase which seems to be separated quite clearly from liquid He I by a phase transition line (see Figure 6). There are two triple points: one on the melting curve at the solid phase boundary and one, the standard lambda point, on the vapor pressure line. There is also a normal liquid–gas critical point; but by previous arguments we should not expect to find any critical points between normal and superfluid liquids since it is now generally recognized that there is an order parameter characterizing the new phase which is peculiar to superfluid helium. We can think of it as a wave function $\psi(\mathbf{r})$, but what that principally means is that it consists of a real part and an imaginary part

$$\psi(\mathbf{r}) = \psi'(\mathbf{r}) + i\psi''(\mathbf{r}), \qquad (15)$$

which in turn is simply saying that the order parameter is a two-dimensional vector. One can also write $\psi(\mathbf{r})$ in the polar form,

$$\psi(\mathbf{r}) = |\psi(\mathbf{r})| \, e^{i\phi(\mathbf{r})}. \qquad (16)$$

The phase $\phi(\mathbf{r})$ is really physically unknowable, since it is just the phase of the wave function which is essentially unmeasurable. Therefore, although the order parameter has this character of being complex or having a phase, the "direction" of the phase is effectively immaterial: I have then, if you like, cylindrical symmetry as shown schematically in Figure 7. Thus the order parameter is a two-dimensional vector and we have a rotational symmetry, the so-called *gauge symmetry*. It transpires that this is of far-reaching significance. A similar situation prevails in the superconductor where again the order parameter has this character of a wave-function. In magnetic systems there is an analogue although it does not apply with quite such complete stringency. The phase or gauge symmetry is a symmetry we do not know how to break in the real world. In many magnetic systems there is an *easy plane* of ordering so that the spin or magnetization vectors prefer to lie in a plane but have no preferred direction in the plane. One often nowadays refers to these as "XY systems", suggesting that there is a spontaneous magnetization given again by a two-dimensional vector. However, one can always break this symmetry by imposing an external magnetic field; in practice, small terms in the Hamiltonian, arising from crystalline fields, etc., usually spoil the perfection of the symmetry even in zero field. If we consider a pure Heisenberg system, where the coupling is a vectorial coupling, of the

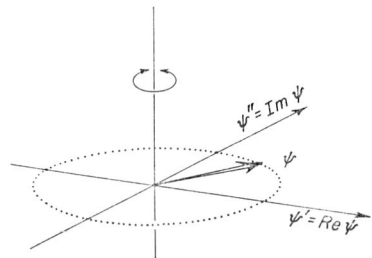

FIGURE 7. Illustration of the cylindrical nature of gauge symmetry.

form $J\mathbf{S}_r \cdot \mathbf{S}_{r'}$, we should think of a three-dimensional vector; in this case, the spontaneous magnetization can point anywhere over a whole sphere and we have a three-dimensional vector, order-parameter. I will mention again the liquid crystals which also have a vector, order-parameter or, perhaps better, a quadripolar order-parameter since normally no sense can be attached to the axis of molecular alignment. As before, if we impose a magnetic field in some particular direction on a ferromagnet, we break the symmetry. When the field is reduced to zero we may be left with some spontaneous magnetization in that direction. That is a broken symmetry, but a broken *continuous* symmetry; and that fact turns out to be of great theoretical significance. Indeed, it limits the type of phase transition we can have. So that is my next major point—the importance of the dimensionality and tensorial character of the order parameter.

V. The Sharpness of a Transition

Another point I want to stress is one mentioned very briefly before; the question of the sharpness of the phase transition. The associated problems are ones which remain confusing to many people. I wish simply to emphasize certain points briefly but strongly. Recall Figure 4 which shows the

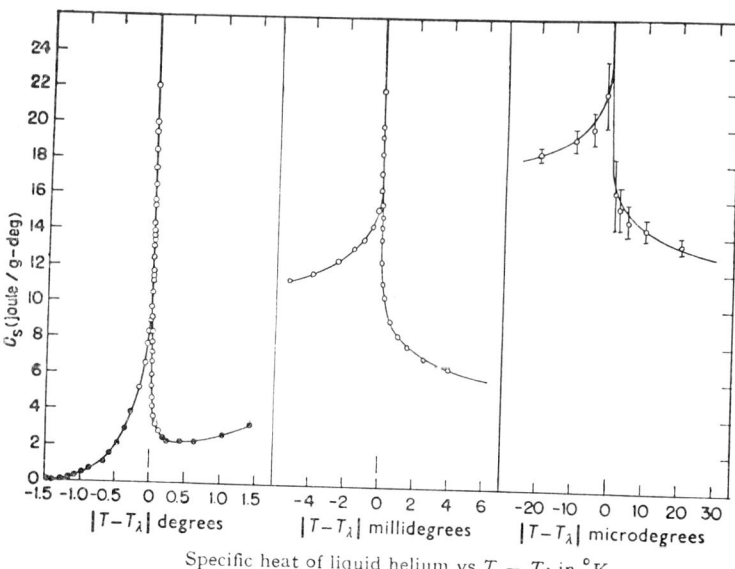

Specific heat of liquid helium vs $T - T_\lambda$ in °K.

FIGURE 8. The specific heat of liquid helium (Fairbank *et al.* 1957).

(sub-lattice) magnetization data for MnF_2 and note the comparatively expanded temperature scale. As you can see, Heller and Benedek (1962) quote the critical temperature T_c to three decimal places—and that is not joking—T_c really is pinned down to about one part in 10^5 at 67.336° K. Thus, the magnetic transition is sharp to a high precision. Similarly, the critical point of carbon dioxide is known to within something like a hundredth of a degree which again means one part in 10^5 or 10^6: evidently, such transitions are extremely sharp when you take precautions. Figure 8 emphasizes this further with the beautiful data originally presented by Fairbank, Buckingham and Kellers (Kellers was the graduate student), of the specific heat of liquid helium showing the lambda transition. The first part of the graph represents the specific heat "anomaly" on a scale of degrees. Multiply the scale by a thousand, and we obtain the middle portion of the graph: we see that the transition is still perfectly sharp on a scale of millidegrees. Lastly, on a scale of microdegrees and you can just begin to see that one may not be quite sure where T_c ($\equiv T_\lambda$) is. Nevertheless, the transition is again pinned down to at least one part in 10^5. The next figure

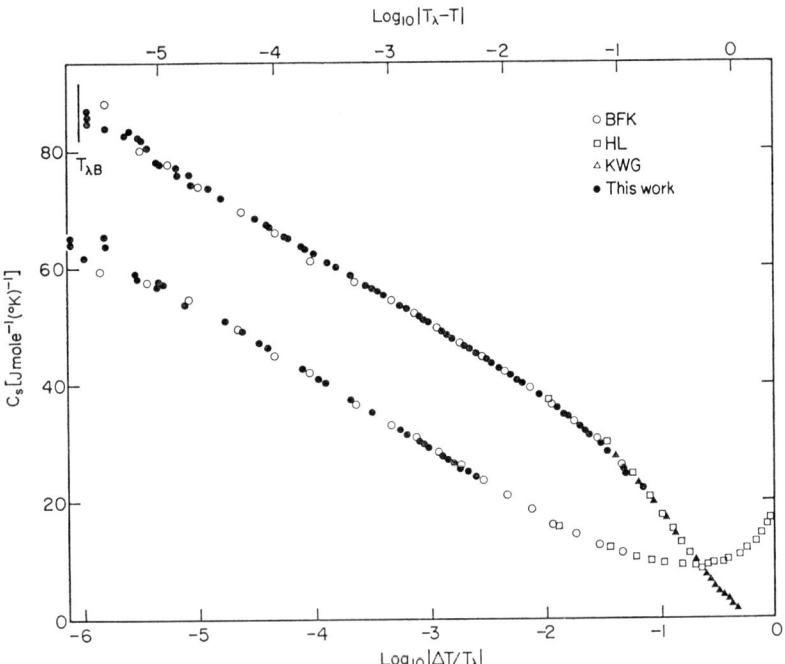

FIGURE 9. The specific heat of liquid helium on a logarithmic temperature scale: $\Delta T = (T - T_\lambda)$, $T_\lambda = T_c$ (Ahlers 1971).

shows more recent data taken by Ahlers, a very talented experimentalist at Bell Telephone Laboratories. The temperature is now plotted on a logarithmic scale with $|T - T_c|/T_c$ ranging from 10^{-1} (or 10% of T_c) to 10^{-6}. As you can see even the last data points above and below T_c indicate that the specific heat is still going up. Hence the critical point (or lambda point) is tied down to one part in a million or better. Indeed, all the evidence suggests that when you have good samples and careful experiments, all transitions should be sharp to this high degree.†

The consequence of this conclusion relates to the general question of how we are going to tackle the theory of critical points. We shall discuss equilibrium systems; I am not going to consider non-equilibrium phenomena. Thus the theory must rest fairly and squarely on the partition function, say for a system of N spins or of N atoms. The free energy is then to be calculated according to the law,

$$F_N(V, T) = - k_B T \ln Z_N(V, T). \tag{17}$$

Of course, the specific heat is given by differentiating this twice with respect to T_c. But note a peculiarity. By definition, the partition function is

$$Z_N(V, T) = \sum_j \exp\left[- E_{N,j}(V)/k_B T \right], \tag{18}$$

where the $E_{N,j}$ ($j = 0, 1, 2, \ldots$) are the overall energy levels of the system, which depend both on the number of particles N and the volume V of the system. The exponential function in (18) is a beautifully smooth function of inverse temperature, not only on the real axis but everywhere in the complex plane: it is an *entire* analytic function. A sum of entire functions is also an entire function.‡ Therefore, the partition function is, as it were, an *infinitely smooth* function of temperature. From this it follows that the free energy is also infinitely smooth. So how can one ever get an abrupt and sharp transition?

An answer to this puzzle is illustrated in Figure 10 which shows the specific heat of an Ising lattice, a very primitive model of a ferromagnet, of *finite* size. If we take a 2×2 Ising lattice, its specific heat is a completely harmless curve; for a 4×4 lattice, a nice rounded peak is beginning to show up; as we go to 8×8, 16×16 and 32×32, the peak becomes increasingly sharp; already for a 64×64 lattice the specific heat rises to more than twice the gas constant. Experimentally, one cannot

---

† One could also mention triple points in place of critical points. The gas–liquid–solid triple point for water is at $-0.0098°C$ and 4.546 mm Hg.

‡ One can show that the infinitude of energy levels does not alter this conclusion.

normally go beyond that by more than a factor of 2 or so. In fact, as N increases, the height of the specific heat peak grows as $\ln N$. This illustrates that to obtain a sharp phase transition we must go to large, ideally infinitely large, systems. One cannot just let N tend to infinity in the formula (17) because the free energy would then "blow up"! Instead one must calculate the free energy *per* spin and take the limit

$$F(v, T) = \lim_{N \to \infty} N^{-1} F_N(V, T), \quad v = \lim_{N \to \infty} V/N. \quad (19)$$

This *limiting* free energy *can* have one or more singularities, which represent completely sharp phase transitions. The singularities "grow" as N increases, becoming perfectly sharp only when "$N = \infty$". If I am seriously interested in phase transitions, I have to pay proper attention to what goes on when I take this limit. Amusingly enough, although this has been appreciated for

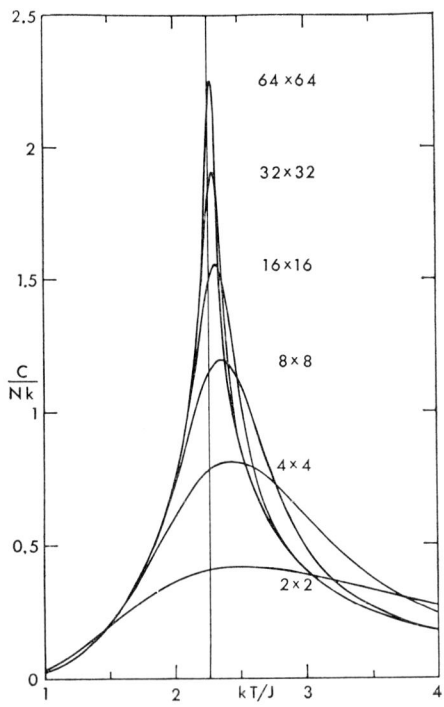

FIGURE 10. Specific heat of finite Ising models: the plots are for $n \times n$ plane square lattices with periodic boundary conditions (Ferdinand and Fisher, 1969).

perhaps 30 or 40 years, it is only in the last four or five years that the existence of this *thermodynamic limit* has been proved satisfactorily from a mathematical point of view (see Fisher, 1964 and Ruelle, 1969). Furthermore, if this limit is to exist, one must say something about the energy levels $E_{N,j}(V)$ or, equivalently, about the types of systems. The thermodynamic limit is not guaranteed to exist in all circumstances; however, when you recognize what happens when it fails to exist you can see that it makes good physical sense.

The next figure gives us Onsager's famous solution for the final limiting specific heat curve of the plane square Ising model, the curve (solid line in Figure 11) goes up to infinity, being logarithmically divergent at a precisely defined T_c, but defined precisely only in the thermodynamic limit. We conclude that, in thinking about experiments and in thinking in detail about theory, we have to bear in mind this way in which a phase transition singularity emerges.

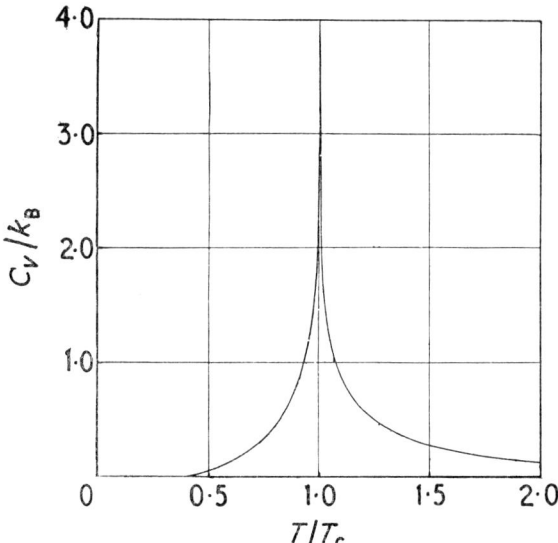

FIGURE 11. The limiting specific heat per spin for an infinite plane square Ising model as calculated by Onsager (1944).

VI. Existence and Non-existence of a Transition

Let us now move on to the question of when a transition can occur and when it cannot. I want to discuss the one-dimensional Ising model: spins with $S_i^z = \pm \frac{1}{2}$ sit on the sites i of a one-dimensional lattice (or linear chain) and the ith spin is coupled to its neighbor, say j, merely by a z–z coupling term $-JS_i^z S_j^z$. For positive J each neighboring pair of spins prefer to point the same way; if they point in opposite directions, they have a larger energy. To solve the one-dimensional Ising model is quite trivial. It involves diagonalizing a 2×2 matrix and Ising in his thesis dissertation managed it, calculated the specific heat and found that nothing happened! Figure 12 displays his result (dashed curve ---). There was no transition, just a smoothly rounded curve; he was very disappointed. In fact, this negative result led Heisenberg to propose his isotropic spin–spin interaction

$$-J\mathbf{S}_i \cdot \mathbf{S}_j = -J(S_i^z S_j^z + S_i^x S_j^x + S_i^y S_j^y),$$

because he thought there was something wrong with the simple Ising interaction which prevented the transition. However, we shall see that this is not true and, ironically, that the Heisenberg interaction in fact makes a transition *more* difficult.

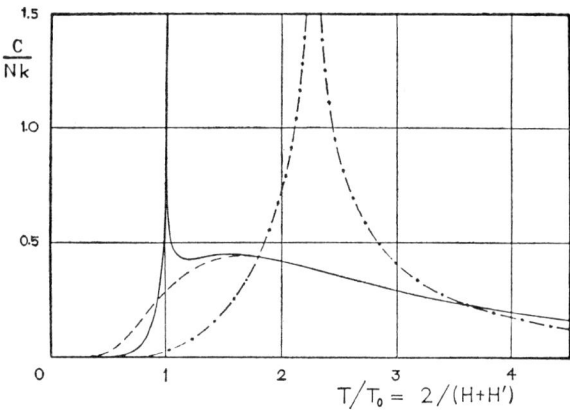

FIGURE 12. Specific heat of a linear Ising chain (dashed line --- merging with solid line): of the symmetric square lattice with $J' = J$ (dot-dash line –·–·–); and of a rectangular Ising lattice with vertical interactions $J' = \frac{1}{100}J$ (solid line) (Onsager, 1944).

Here we see our first instance of a change of dimension entering the picture: the two-dimensional Ising model has a beautiful transition with a logarithmic specific heat; the one-dimensional model has none. Can we not find some way of going over from the one to the other? Consider a square lattice array of spins and let us couple together spins in the same horizontal row by an interaction J; then we have simply a host of independent one-dimensional chains. But now let us couple vertically neighboring spins by a second interaction J'. If J' becomes smaller and is finally reduced to zero, the system becomes one-dimensional; but if J' is non-zero, the system suddenly becomes two-dimensional. What happens to the transition? Onsager took care of this and in Figure 12 you see what happens: the large anomaly shown by a dot–dash curve is just the specific heat, which we saw before, of the symmetric square Ising lattice where the two interactions are equal; the solid curve gives the specific heat when the interaction J' has been cut down by a factor of 100. The transition then takes place at a lower temperature; but it is surprising that the transition temperature is reduced only by about 40%. Mark that for future reference: a 40% reduction follows a one-hundredth reduction in J'. The specific heat anomaly is smaller but it is still logarithmic, and it is still infinitely sharp. As you see, it just sits on top of, and modifies slightly, the linear chain result. If J' is made weaker still, say by another factor of 100, the transition temperature will drop further and so will the magnitude of the anomaly which, however, remains a logarithmic spike sitting on the linear chain result. Finally, what happens is that the phase transition "bows out" through zero temperature. Conversely, if I start with the uncoupled chains and switch the two-dimensionality on, the phase transition grows quite rapidly out from zero temperature.

This is my first illustration of the way in which a phase transition can be influenced by dimensionality. In the next figure you see that this sort of a system is not just a theoretician's fiction. There are some nice materials: the one illustrated in Figure 13 is copper tetramine sulphate, $Cu(NH_3)_4 SO_4 \cdot H_2O$. The data were obtained by Haseda and Miedema (1961) in the Netherlands but Stout in Chicago had previously studied a similar system. Most of the experimental curve in Figure 13 (solid line) looks just like the

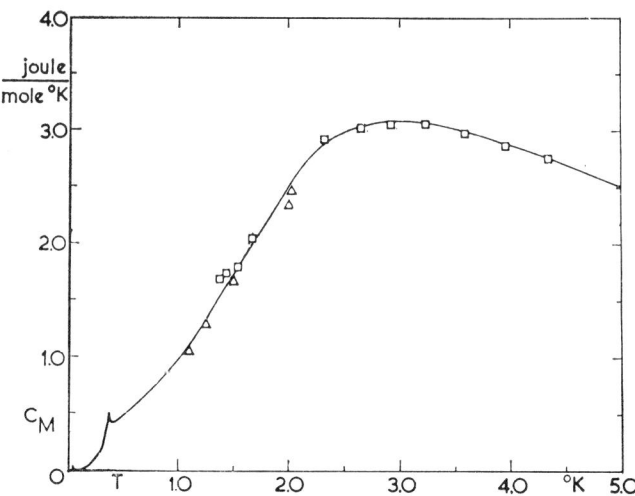

FIGURE 13. The specific heat of $Cu(NH_3)_4SO_4 \cdot H_2O$ (data points and solid curve) (Haseda and Miedema, 1961).

specific heat of a linear chain and indeed one can tell from the known structure of the material that the magnetic copper ions are arranged mainly in linear chains. But these chains must be coupled together to some extent and so eventually, at low temperature, the crystal orders. Below T_c the chains are aligned together three-dimensionally. The ordering is associated with a little specific heat anomaly which is shown in more detail in Figure 14.

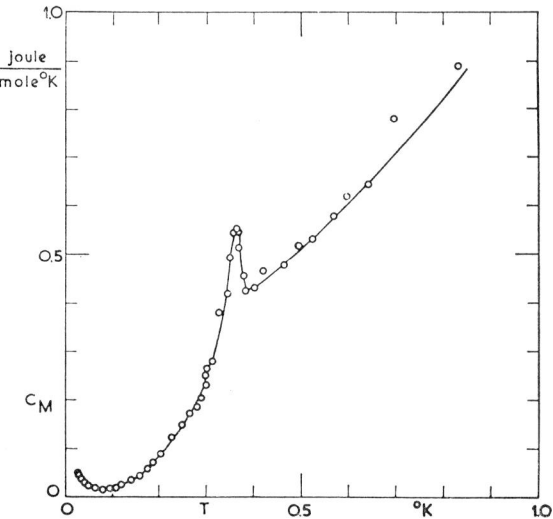

FIGURE 14. Detail of the specific heat anomaly in $Cu(NH_3)_4SO_4 \cdot H_2O$ (Haseda and Miedema, 1961).

The similarity to the results found by Onsager for Ising chains weakly coupled together is striking. If one could pull the chains of copper ions further apart by forcing some more non-magnetic atoms into the crystal between them, the critical temperature would drop and eventually go below the millidegree region; we would still be chasing it at Cornell and at Argonne and at other low temperature laboratories to see where it had gone! Since the crystal would really remain three-dimensional, one of the conclusions we obviously reach is that eventually we would catch up with it!

VII. Transitions in One Dimension

These results concerning lack of transitions in one dimension, are well known. However, I would like to review the theoretical argument for you since it does not follow that things which are "well known" are always correct. I hope to indicate the limitations of the argument but at the same time show how they do reveal clearly the influence of dimensionality. The argument is most simply put, following Landau, by considering a one-dimensional spin system and supposing that it is fully ordered so that all the spins point 'up'. Now if we want to disorder this state, we have to make a break in the spin alignment and from some point along the chain have the spins point 'down'. This break in the alignment will cost some energy, essentially just the energy lost across the "interface" between the 'up' domain and the 'down' domain. Let us suppose that energy is W. Is this disordering process likely to happen by itself? We go back to our elementary thermodynamics and say if the associated free energy change is less than zero, it will happen spontaneously; if it is greater than zero, it will not. The free energy is $F = U - TS$ where U is the internal energy and S is the entropy. The energy change is simply $\Delta U = W$ but if the system is N spins long, there is also an entropy change for the simple reason that this interface can be put in a large number of places. The entropy is always the logarithm of the number of configurations and hence the entropy change is just $\Delta S = k_B \ln N$. Thus at any finite temperature, the free energy change

$$\Delta F = W - k_B T \ln N \qquad (20)$$

will be negative in a sufficiently large system. Therefore, it will always pay thermodynamically to break up a one-dimensionally ordered system. The argument also shows, correctly, that if I have a finite system (i.e. a finite N) then at a low enough temperature, it will effectively go into one single aligned domain. But the system will not go into that domain sharply. If one works out the probability of having an interface or break at a given point, it is essentially $e^{-\beta W}/(1 + e^{-\beta W})$, which is a perfectly smooth function of $\beta = 1/k_B T$. The probability of a break does tend to zero as the temperature decreases, but it is always non-zero for $T > 0$. That concludes the argument against one-dimensional transitions.

Next, let us go a little further and recall an argument originally due to Peierls (1936) concerning the *occurrence* of a transition in two dimensions. Unfortunately, when he wrote it out he made a slip in the mathematical logic, although the physics was good. It took some thirty years before R. B. Griffiths put the logic right. In the meantime a number of people repeated his argument without apparently noticing the lack of logic. I take a little credit for noticing the gap in the logic, but not for being able to put it right! Griffiths (1964) and, independently, Dobrushin (1965) in Russia gave the first rigorous proofs. But I will present the rough arguments and never mind the gap in logic, the physics is correct! Consider then a fully aligned $L \times L$ plane, square lattice so that we have $N = L^2$, and ask: "What is the free energy change if somehow or other I break the ordered lattice into an 'up' domain and a 'down' domain?" Thus, from some point on the edge of the system let me start to draw a domain wall (see Figure 15).

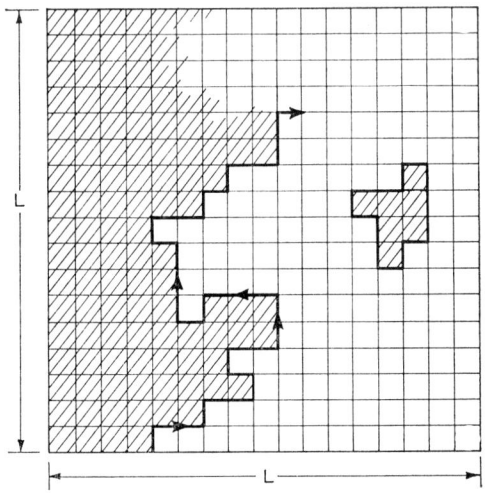

FIGURE 15. Illustrating the stepwise construction of a domain wall across a two-dimensional square lattice, and a small, finite microdomain or "bubble".

You can see that as I draw the domain wall, I can gain quite a lot of entropy by "wiggling around". So allow me a wall of length, let us say $s = fL$, where $f > 1$. (The factor f must exceed unity to get across the lattice at all). The total energy change involved is then, $\Delta U = sW = fLW$. What about the entropy? How many configurations of the wall are there? I can choose any point along the bottom or, say, left-hand edge of the lattice to start from. So the number of possible configurations is proportional to $2L$. Then, since we have a square lattice, I can at each step go straight ahead, turn left, or turn right. I might run out at the edge or recross my path in doing that, but at least I will not underestimate the entropy if I say that the number of configurations is something like $2L(3)^{fL}$. The entropy change thus satisfies

$$\Delta S \leqslant k_B \ln [2L(3)^{fL}] = fLk_B \ln 3 + k_B \ln 2L. \qquad (21)$$

The logarithmic term $k_B \ln 2L \sim k_B \ln N$ was actually the only one that occurred previously, see equation (20): in one dimension it gives the dominant contribution, but now you see it is negligible compared with $fLk_B \ln 3$. The free energy change needed to disorder the lattice is hence estimated by

$$\Delta F \geqslant fN^{\frac{1}{2}}(W - k_B T \ln 3) - \tfrac{1}{2}k_B T \ln 4N. \qquad (22)$$

The situation is clearly different from that in one dimension. For a large system we can ignore the second term; then at a low enough temperature—specifically if $T \leqslant T_c' = W/k_B \ln 3$—the free energy change in putting *any* such wall right across the lattice is positive. Not only is it positive, but it is very large and positive; and as N goes to infinity it diverges. We conclude that the aligned state is thermodynamically stable against break-up into disordered domains; a true phase transition should occur. Above the temperature T_c' or, at any rate sufficiently far above it, we can see that the entropy should win and destroy the ordering. Thus the argument gives us both an estimate of the critical point (actually a lower bound) and a conviction that, because of its dimensionality, the system will at a low enough temperature, go into an ordered state which cannot be broken up. We could, of course, blow small, finite "bubbles" or microdomains as indicated in Figure 15. Such microdomains occur with definite non-zero probability and give the spontaneous magnetization its temperature dependence. However, below T_c the chance of having a barrier right across the lattice becomes vanishingly small; it is roughly proportional to

$$L\,e^{-\beta J L W'}/(1 + L\,e^{-\beta J L W'}),$$

where

$$W' = W - k_B T \ln 3, \qquad L = N^{\frac{1}{2}}. \tag{23}$$

As N goes to infinity this goes to zero very rapidly, provided $W' > 0$ or $T < T_c'$.

As I have mentioned, this argument has in recent years been made quite rigorous; we now have rigorous proofs, applicable in any number of dimensions greater than unity, that these simple systems (Ising models) have a phase transition. Of course, this result was expected but it is important to check our physical understanding with good mathematics; happily, the mathematics uses just this physical argument.

Now a word of caution! The next two figures (Figures 16 and 17) show data for biochemical rather than solid state systems. They represent the so-called "melting" of long polymer molecules in solution. Under suitable circumstances a polymer chain will coil up into a helix or spring-like "ordered" state: this corresponds to a crystal. If one heats a solution of such molecules or alters say the pH (as in Figure 16), the coiled-up helices "melt" into disordered "random chains". One sees this by following, for example, the optical rotation or the ultraviolet absorption, etc. A rather definite transition can be seen in Figure 16; evidently however it is spread out with quite a smooth transition region and no unique melting point. When people first saw this sort of behavior they were pleased: "Polymer chains are essentially one-dimensional systems and hence they shouldn't melt sharply." However, Figure 17 indicates that one should not jump to conclusions too rapidly! In a closely similar, albeit synthetic, polymeric system the melting (this time as a function of temperature) is very sharp. Notice that the degrees are centigrade not absolute. Very abruptly, within an interval of 0.1° or less, the chains suddenly melt or, if one cools the solution, suddenly freeze into helical form. The transition appears to be of order higher than first order but it is sharp to three parts in 10^4—not yet one part in 10^5 or 10^6, as we saw previously, but these experiments have not been pushed. (Note that the relevant dimensionless figure of precision is $\Delta T/T_c$ where T_c is the absolute, thermodynamic temperature and ΔT is the uncertainty in the critical point).

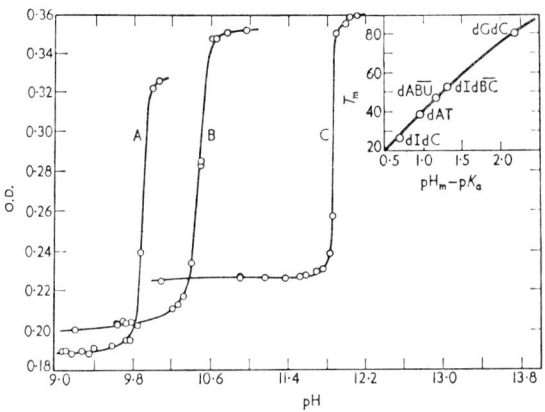

FIGURE 16. The "melting" of polynucleotides as a function of pH, as observed by ultraviolet absorption. Variation of temperature would yield similar results.

Let me explain what we can learn from this example. In discussing one-dimensional systems and in deriving and analysing the formula (20) I made one key assumption; namely, that when an ordered domain is broken by inserting an interface or wall it costs energy W, where W is *independent* of how long the system is. That is likely to be true if there are only short-range interactions along the chain, but if one has long range interactions in the chain, it no longer need be so. In that case, W will depend on N: then if $W_N \ln N \to \infty$ as $N \to \infty$, the previous argument fails and, just as in two dimensions, we would conclude that an ordered state *could* be stable. What are the long range interactions in a polymer chain? Although a polymer molecule is a one-dimensional chain, it can wiggle around in three dimensions. In this way it can gain entropy which, because of geometrical considerations and the formation of loops, has essentially long range (logarithmic) terms in it. The long range "forces" thus enter in a rather hidden way. In more familiar physical systems this question of long range force is somewhat academic, but not entirely so. In fact Dyson, in the last year, has proven that if one has an Ising chain with spin–spin interactions which fall off with the distance as

$$J_{ij} \sim 1/r_{ij}^{1+\sigma}, \tag{24}$$

then there *will* be a transition if this fall-off is slow enough. The criterion for a transition is $\sigma < 1$; one must have $\sigma > 0$ if the model is to make thermodynamic sense. If, on the other hand, $\sigma > 1$, then it was proved by Ruelle somewhat earlier that there can be no transition. A one-dimensional system will thus not have a phase transition if the interaction falls off sufficiently rapidly. It is interesting that these rigorous results were anticipated by

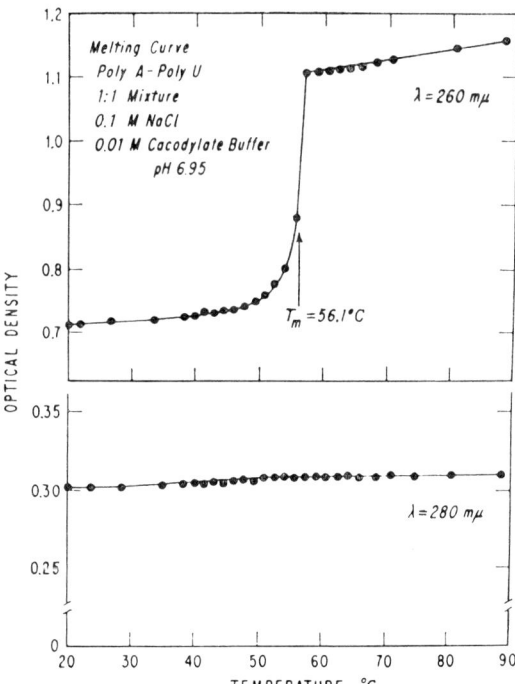

FIGURE 17. A sharp melting transition in a synthetic polynucleotide; optical density versus temperature (Stevens and Felsenfeld, 1964).

extending the previous physical arguments: basically one must just calculate W_N and see how it compares with the entropy.†

What happens exactly *on* the borderline $\sigma = 1$, i.e. when $J_{ij} \sim 1/r_{ij}^2$? I do not know. Dr. C. J. Thompson has some interesting views on the problem and others among us in the field have also made intelligent guesses.‡ P. W. Anderson has pointed out that the question is closely related to the behavior of a special case of the Kondo problem which he has recently discussed in detail. He concludes that there is a transition, although the arguments are not yet very transparent mathematically. At this moment the problem seems to resist fully rigorous analysis. Perhaps you might regard it as only a theoretician's plaything but I think the connection with the Kondo problem gives one a feel for the depth and implications of some of the questions arising in the study of phase transitions.

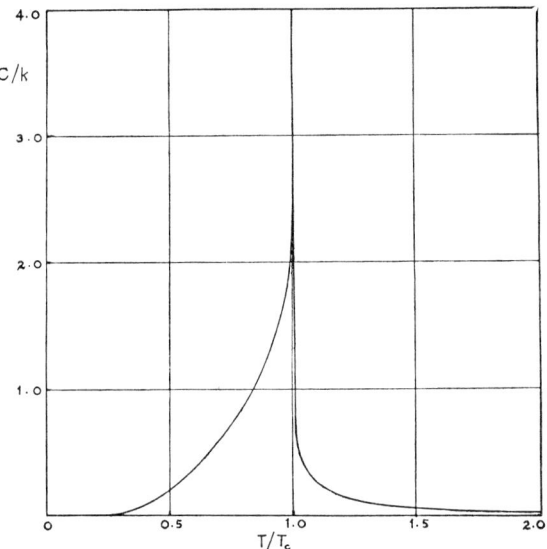

FIGURE 18. The specific heat of the three-dimensional Ising model for a face centred cubic lattice (after Domb, 1960).

† The argument yields $W_N = \sum_{l=1}^{\infty} (l-1)[J_{0,l} - J_{0,l+N}]$ and predicts "no transition" if, and only if, $W_c = \lim_{N \to \infty} W_N/\ln N = 0$.

‡ For $J_{0,l} = l^{-2} (\ln l)^\alpha (\ln \ln l)^\beta$ the simple argument predicts a transition for all β if $\alpha > 0$ but only for $\beta \geq 0$ if $\alpha = 0$.

VIII. Dimensionality and Critical Behaviour

At this juncture I shall draw attention to an intriguing, but less well-understood, way in which dimensionality affects phase transitions. Consider magnetic systems in which the dimensionality is sufficiently great to insure a phase transition and look at the critical point behavior and, especially, the critical exponents. Figure 18 shows the specific heat curve of a three-dimensional Ising model. Somebody may have told you that no one has solved the three-dimensional Ising model. Analytically, that is true but for numerical purposes it has been solved some time ago. Comparison with the curve shown before for three-dimensional Ising models (Figure 11) reveals an interesting difference: in two dimensions the specific heat rises symmetrically on each side of the critical temperature T_c. What we see in three-dimensions is more characteristic of the real lambda transition shown in Figure 8; there is much more specific heat on the low temperature side than on the high temperature side of T_c. That asymmetry seems to be a characteristic change on going from two dimensions to three dimensions.

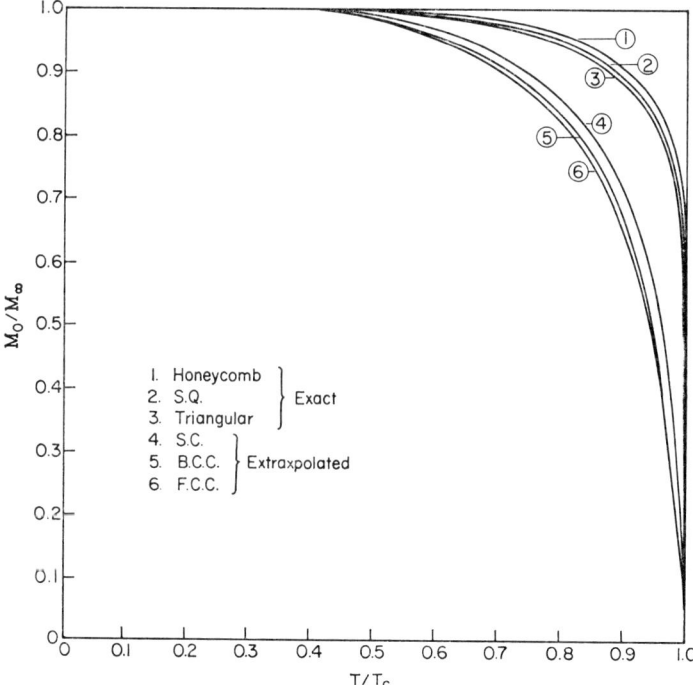

FIGURE 19. Spontaneous magnetization curves for various two- and three-dimensional Ising models normalized by T_c (after Domb, 1960).

There are similar changes in other quantities: Figure 19 shows the behavior of the spontaneous magnetization. The first three curves are for two-dimensional Ising models of various kinds: honeycomb, square and triangular. They all come down very steeply with a critical exponent of exactly $\beta = \frac{1}{8} = 0.125$. [Recall the definition (3).] The curves in the second set for the three-dimensional lattices are much more rounded which means they have a larger exponent. In fact, the three-dimensional exponent is $\beta \simeq 0.313$ which is very close to $\frac{5}{16}$. This value is close to, but slightly lower than the observed values of $\beta \simeq 0.33$ to 0.36. You should remember the contrast between these values because later on I will present experiments which yield values of β in the vicinity of $\beta = 0.1$ to 0.2 rather than in the region of $\beta \simeq \frac{1}{3}$—experiments that are perhaps pseudo-two-dimensional! The specific heat and the spontaneous magnetization suffice to make my point although I could show you much more evidence. All critical point behavior seems to be very strongly dependent on the dimensionality. We have some rough ideas of why this is so. Thus one can extend the arguments involving interfaces and domain walls which we discussed before, and they lead in that direction. There have been a number of developments from that; but in the main they just give hints. We do not have a really sound theory as to why the exponents change as they do. (But see the Postscript at the end of this article.)

IX. Absence of Ordering in Two-dimensional Isotropic Systems

Having discussed one-dimensional systems at some length and shown that, at least for short range forces, there will be no transition, let us turn to two dimensions. We know that the two-dimensional Ising model, which is quite a good model of many systems, binary alloys in particular, *does* have a transition. What about the Heisenberg model mentioned above? This is certainly a better model for a magnet because it has spins that are not just coupled through the z components as in an Ising model, but rather are coupled together isotropically through all components. Does it order in two dimensions? Here we must recall the point I stressed before: namely, the order parameter in the Heisenberg model has a different character. The Ising model order parameter is just "up-or-down," a trivial discrete one-dimensional order variable. In the full Heisenberg model the order parameter is a three-dimensional vector **S**—we have a continuous symmetry, the system is "isotropic" in spin space. Even in an XY model, the order parameter is a two-dimensional vector. The history of this is interesting; perhaps it shows the way theoretical physics often goes. Back in the early days, Bloch and Bethe made a serious attack on the Heisenberg model and were able to find the ground state of the one-dimensional chain. We do not know to this day, in strict mathematical terms, whether that ground state is ordered or not, although we can be rather sure that it is not. Of course Bloch and Bethe went on to look at two and three dimensions. They assumed that the system is ordered and then found that there are spin waves—some of you will have seen the calculations. If one sticks to three dimensions, everything is fine: at low temperatures there are few spin waves; as T increases the population of spin waves grows and they interact; eventually the theory breaks down; it is a low-temperature theory. If you follow the arguments in two dimensions, as Bloch originally did, you find that at any finite temperature, there are so many spin waves wanting to be around, that the whole calculation blows up in your face. Now, if you were really an honest theoretician, you would say: "My scheme of calculation has broken down and I am going to keep discreetly quiet." But, of course, not many of us are that scrupulous, and in fact, almost nobody said anything of that sort. On the contrary, the response tended to be: "My calculations are pretty good. This just tells me that there cannot be an ordered state in two dimensions. I put the blame on Nature!" It all turns on the divergence of the simple d-dimensional momentum-space integral,

$$I_d(\lambda) = \int_{|\mathbf{k}|<Q} d^d\mathbf{k}/(k^2 + \lambda),$$
$$= C_d \int_0^Q k^{d-1}\,dk/(k^2 + \lambda), \quad (25)$$

as the parameter λ tends to zero. (The constant C_d involves only numerical factors.) One finds

$$I_d(0) < \infty \quad \text{for} \quad d \geq 3,$$

but

$$I_2(\lambda) \sim \ln \lambda^{-1} \to \infty, \quad (\lambda \to 0),$$
$$I_1(\lambda) \sim \lambda^{-\frac{1}{2}} \to \infty, \quad (\lambda \to 0). \quad (26)$$

For many years people seem to have believed this "folk theorem" regarding the absence of a phase transition in a two-dimensional Heisenberg model although it can scarcely be regarded as a convincing proof. But every now and then it came under attack. In particular, one can try to calculate the "initial" or zero-field susceptibility, $\chi_0(T)$, of a two-dimensional Heisenberg model not at low temperatures, but at high temperatures. One discovers a power series in inverse temperature; in fact, with care and hard work one can obtain quite a few terms of the series; G. S. Rushbrooke was one of the first people to develop this. As Rushbrooke pointed out, when you actually examine these series they look much the same in two dimensions as in three. In three dimensions one is convinced that $\chi_0(T)$ should blow up, giving one an infinite susceptibility at the Curie point $T = T_c$. The divergence is not quite that given by the Curie–Weiss Law since again one of the critical exponents comes in (this one is called γ); but still $\chi_0(T)$ should diverge and the series expansion coefficients do indicate this clearly. Indeed, one can use them to estimate the critical point rather accurately, as Rushbrooke did. When one makes a similar analysis of the two-dimensional series one finds they are more irregular but they also look as though the susceptibility diverges at some non-zero critical temperature. Rushbrooke pointed this out but somewhat apologetically and without great conviction because, of course, everybody "knew" that the two-dimensional Heisenberg model should not have a phase transition. Some years later, Stanley and Kaplan (1966) obtained a somewhat longer series and a more regular series since it was for high spin. Apparently they did not believe the folklore so strongly because they made the rather bold conjecture that the two-dimensional Heisenberg model really *did* have a phase transition. About the same time Hohenberg (1967) at Bell Telephone Laboratories, and Mermin and Wagner (1966) at Cornell (Hohenberg thinking primarily about the superfluid case) showed rigorously that, at least by some of my tokens of a phase transition, these "isotropic" two-dimensional systems could not and did not have a phase transition. The arguments of Mermin and Wagner were completely rigorous and I would like to summarize some of the facts we now know. Everything still turns on the divergence of the integral $I_d(\lambda)$ defined in (25). In fact, the mathematical source pinpointed by Bloch can, in some ways, be said to be absolutely correct. Keeping in mind some of the criteria of order which we discussed, first notice that what Mermin and Wagner actually established was that you could not have any spontaneous magnetization. Specifically, they obtained a general upper bound on the magnetization in a field, by using an inequality that the famous Russian theoretician Bogoliubov (1962) had produced. Originally, this inequality looked most unhelpful and Bogoliubov buried it in a little-read East German journal of physics. But it turned out to be extremely useful— by useful, I mean, at very least, that many people, including myself, have been earning their living on it every since! Mermin and Wagner showed that Bogoliubov's inequality could be manipulated to yield

$$[M(H, T)]^2 \leqslant c/T I_d(c'H), \qquad (27)$$

where c and c' are, effectively, constants. Now as $H \to 0$ we see from (26) that

I_d diverges if $d \leqslant 2$ and thence obtain

$$M(H, T) \leqslant c_2/T^{1/2} |\ln (c'H)|^{1/2}, \quad (d = 2),$$

$$\leqslant c_1 H^{1/3}/T^{2/3}, \quad (d = 1). \quad (28)$$

In both one and two dimensions the bound on $M(H, T)$ goes to zero with H, at all non-zero temperatures. Hence the spontaneous magnetization must be identically zero. [Recall the definition (2).] The proof works as well for the XY model as for the full Heisenberg model—all one needs is one axis of rotational symmetry in spin space.

These results, although gratifying, do not quite conflict with what Rushbrooke, and Stanley and Kaplan were saying. The spontaneous magnetization tells us what happens *below* T_c but the divergence of the susceptibility *above* T_c as one goes to the critical point is something else. In fact, the divergence of $\chi_0(T)$ provides a good criterion for a (higher order) phase transition although I have not discussed this. Somehow or other, in all familiar, real ferromagnets, once the susceptibility has gone to infinity on reducing the temperature T, the system becomes spontaneously magnetized; i.e. the criteria $M_0(T) \to 0$ and $\chi_0(T) \to \infty$ agree. In two dimensions M_0 must vanish below "T_c" but, following the suggestion of Stanley and Kaplan, might not $\chi_0(T)$ diverge as $T \to T_c+$ and then *remain infinite* for all $T < T_c$? This suggestion is not as peculiar as it first sounds when we notice that the bounds $M^>(H, T)$ on $M(H, T)$ in (28) have themselves the property that $\chi^>(H) = (\partial M^>/\partial H)_T$ diverges as $H \to 0$.† An infinite initial susceptibility thus merely means that the magnetic (M, H) isotherm leaves the origin with an infinite slope—the isotherm may, and will, have a finite slope for non-zero H. Still the answer to the question is that we have not, as yet, proved anything about the initial susceptibility although I and my colleagues at Cornell have tried to make progress towards it. Let me tell you briefly about some of the more recent developments.

First of all, I should say that in some of these matters strong emotions are aroused! One of the paradoxes is that this proof, when suitably rephrased, rules out the expected "off-diagonal" order in superconductors and superfluids. All the people who had been happily passing supercurrents through one-dimensional wires, suddenly felt the ground cut out from under them, because the standard theories—the Landau and the BCS theory—really involve the assumption that there *is* such a spontaneous order; but in any fully satisfactory theory of one-dimensional wires, there cannot be! In fact, these proofs have raised, or re-emphasized, the question of why something

---

† This is also a property of the spin wave approximation for $M(H, T)$ even when $d = 3$.

is a super conductor or a superfluid. The true answer to that is: "It's not just a question of equilibrium. You cannot simply say it's because of the presence of spontaneous order." Thus, as I have said there is *no* spontaneous order in one and two dimensions in the sense that people always used to say there was although, of course, we know that superconducting wires do very well! Actually when you push the analysis a little further you find that they only do so well because we human beings have a finite and small lifetime compared to their finite, but normally astronomically great, lifetime. Some theoreticians who did not want to think along those lines mounted a fresh attack by saying, "Well, the proofs presented assume that the system is *strictly* one-dimensional or strictly two-dimensional. Who ever saw a strictly two-dimensional superfluid or superconductor? Real systems always have a finite thickness of finite cross-section and the proofs must break down, especially if you do not impose artificial periodic boundary conditions." Now, in truth, this objection is physical nonsense. But since it was such recurrent nonsense, some of us recently took the trouble to dispose of it rigorously (Chester *et al.*, 1969). One can indeed discuss a system of finite cross-section or a slab of finite thickness and all our theorems still hold.

In tackling questions about an infinite initial susceptibility the next step is to look at the short-range order, i.e. the pair correlation function $\sigma(\mathbf{r}, \mathbf{r}')$ defined in (5). From $\sigma(\mathbf{r}, \mathbf{r}')$ one can obtain the susceptibility simply by summing on **r** according to

$$\chi_0(T) = (m^2/k_B T) \sum_\mathbf{r} \sigma(\mathbf{r}; T). \tag{29}$$

Where $H = 0$, m is a suitable magnetic moment and the thermodynamic limit has been taken so that (9) applies exactly. In the absence of long-range order, which might be expected to follow from the conclusion $M_0 \equiv 0$ although it does not necessarily do so, we have $\sigma(\mathbf{r}) \to 0$ as $r \to \infty$. We might expect this to mean that χ_0 was finite, but a moment's reflection reveals that if $\sigma(\mathbf{r})$ decays to zero *sufficiently slowly* the sum can diverge so that $\chi_0 \equiv \infty$. This indicates the great interest that attaches to the decay of $\sigma(\mathbf{r})$ in a two-dimensional isotropic system.

Another reason for studying $\sigma(\mathbf{r})$ is that the definition of spontaneous order employed by Mermin and Wagner (and, implicitly, by Hohenberg) uses, as I have explained, a symmetry breaking field which is allowed to approach zero. For ferromagnets this field is just H but for superfluids or superconductors nobody knows how to produce the appropriate analogue of H! Thus one cannot "let this field go to zero" because in the real world it is apparently always zero. Hence one would like to prove something about $\sigma(\mathbf{r})$ for large **r** in strictly zero field. Recently, by developing the earlier arguments, David Jasnow and I have been able to make some progress in these directions (Jasnow and Fisher, 1971). Specifically, we examine the mean-square magnetization per spin, as defined earlier in (7), for some subset of spins Γ. (We may actually allow Γ to be the whole system Ω.) If, as the subdomain Γ became larger and larger, I found that the mean square magnetization remained finite, then despite Mermin and Wagner's result, I would say the system is displaying some sort of long-range order and I would hence expect a true phase transition to occur at some temperature.

It turns out that one can exclude this particular possibility quite rigorously. To state our results, consider a two-dimensional spin system of thickness D (i.e. D layers of spins) and let the linear dimensions of the subdomain Γ be L_Γ (one may suppose Γ is an $L_\Gamma \times L_\Gamma \times D$ "slice"). Then we establish a bound of the form

$$M_\sigma^2(\Gamma) = \left\langle \left[(N_\Gamma)^{-1} \sum_{\mathbf{r} \subset \Gamma} S_\mathbf{r} \right]^2 \right\rangle = (N_\Gamma)^{-2} \sum_{\mathbf{r} \subset \Gamma} \sum_{\mathbf{r}' \subset \Gamma} \sigma(\mathbf{r}, \mathbf{r}')$$

$$\leqslant (b_2 D J S^2 / k_B T) / \ln(b_\Gamma' L_\Gamma), \qquad (d=2) \qquad (30)$$

where S is the spin magnitude, J is the (average) exchange interaction and b_2 and b' are constants. Now, as Γ becomes large $L_\Gamma \to \infty$ and the bound decreases, admittedly rather slowly, to zero. In other words, in an infinite two-dimensional system the mean square magnetization per spin always vanishes. Notice how the bound depends on the thickness D—as is to be expected, it weakens the inequality but does *not* change its form. For a one-dimensional system we can do a little better: if the cross-sectional area is A (i.e. A parallel chains of spins) we find a bound of the form

$$M_\sigma^2(\Gamma) \leqslant (b_1 A J S^2 / a k_B T)^{\frac{1}{2}} / L_\Gamma^{\frac{1}{2}}, \qquad (d=1), \qquad (31)$$

where a is the lattice spacing. In fact, in certain circumstances, and at the cost of some mild assumptions, we can go on from these results to bounds on the correlation function itself. These are of the form

$$\sigma(\mathbf{r}) \leqslant a_2 / \ln r, \qquad (d=2),$$
$$\leqslant a_1 / r^{\frac{1}{2}}, \qquad (d=1), \qquad (32)$$

where a_1 and a_2 are constants which depend on T, and on D and A. Both of these bounds go to zero as $r \to \infty$ so we have proved that the short long-range order vanishes for all $T > 0$. One can even prove that the long long-range order (which I have not discussed) vanishes identically.† Unfortunately, we have been unable to rule out an infinite initial susceptibility. As can be seen from (32), although $\sigma(\mathbf{r})$ must decay to zero it could do so sufficiently slowly (both for $d = 1$ and 2) that the sum (29) for χ_0 *could* diverge.‡ So here

---

† The mathematically inclined reader might wonder if there are delicate questions with regards to shape and so on, in discussing the large size limits in one and more dimensions. Actually, these fine points are under rather good control nowadays. The important point is the behaviour of the volume-to-surface ratio. What we do theoretically is to consider a system which if it is, say, "two-dimensional", can be confined between two parallel planes, but otherwise it does not matter what detailed shape you choose. Then to define the subdomain Γ involved in (30) and (31), one cuts out some sort of "slice" from it—with a cookie cutter, let us say. What basically has to be done for such a system of finite thickness, is to show that it is not too difficult to disorder the spins inside Γ whatever the spins outside Γ are doing. In fact, there is rather a good heuristic argument (which I did not have time to present) which gives the right answers. Just as before, one calculates the free energy involved in inverting the spins inside Γ while the rest of the system remains as an 'up' domain. Now if I wish to break the overall spin alignment in this way, I should not make an abrupt break. Rather, as Bloch pointed out, I should make a 'Bloch wall'. In other words, I should twist the spins rather gently from 'up' to 'down' over a width b. Then, if I go back and calculate the energy W, I find it depends on b. In fact, in a one-dimensional system, I can make the Bloch wall energy as low as I wish by choosing b large enough. Thus, once again the system can never order. When I build myself a Bloch wall in two dimensions—say, a corridor of width b around a circular 'slice' domain—I find that if I make the corridor wide enough, the energy of the wall is essentially constant, independent of how big the domain of overturned spins is. This contrasts with anisotropic, Ising-like systems where, in two or more dimensions, a larger domain demands a larger wall energy which increases without bound as the domain becomes macroscopic. [See the discussion leading to Eqn. (22)]. In an isotropic system there are lots of possible fluctuations and only a constant energy opposes even the largest fluctuation, so, eventually, the system will fluctuate and destroy any long-range order. Indeed, one can actually understand the dimensionless combinations that enter the rigorous inequalities (30) and (31). The factor DJS^2/k_BT is simply the ratio of the two-dimensional Bloch wall energy to the mean thermal energy—obviously the wall energy is proportional to the thickness D. Similarly, in the one-dimensional case, the wall energy is proportional to the cross section A, and inversely proportional to the width b, which is, at most, proportional to L_Γ. The ratio to the thermal energy is thus $AJS^2aL_\Gamma k_BT$, which is just what enters (31). Thus, the heuristic argument works well although, of course, one can never be too sure about things until one has the argument under proper mathematical control. That we have now achieved and, to answer again the particular question raised, the only shape criterion one need employ concerns the asymptotic ratio of the surface to the volume.

---

‡ Indeed there is an appealing but approximate spin-wave-type argument which says that in two dimensions $\sigma(\mathbf{r})$ should decay as l/r^{cT}, which form has such features: it implies a diverging susceptibility below some T_c but no long-range order. However, this prediction for $\sigma(\mathbf{r})$ also has some completely wrong features! In particular, there are rigorous arguments for the Heisenberg model, and for most such models, which tell one that at all high enough temperatures one must have an exponentially decaying correlation function—not just an inverse power law. Therefore, the approximate argument must break down somewhere. Thus, as with the original spin wave arguments, it is not clear that we are on the right track. The nice thing about the heuristic interface arguments is that they contain properly the needs of their own breakdown: for example, they establish the stability of the ordered phase only *below* some critical temperature.

is one of the interesting open questions: and when you have an open theoretical question, of course, you should look at experiment.

X. Two Dimensions in the Real World?

I want to discuss some recent experiments on a material that really does seem to give two dimensions in the real three-dimensional world. Figure 20 shows the crystal structure of the material K_2NiF_4 and of a related substance $KNiF_3$ with only one potassium ion for each nickel ion. The latter is a fairly uninteresting system: the nickels do the magnetic work for us and the crystal is an orthodox antiferromagnet as can be seen by noting the spin alignment indicated in the figure. By doing a little chemistry, one can insert a lot more potassium (and fluorine) into the $KNiF_3$ lattice: what happens is that the layers of nickel ions (black circles in the figure) are pushed further apart, as is evident in Figure 20. You might say that the structure is

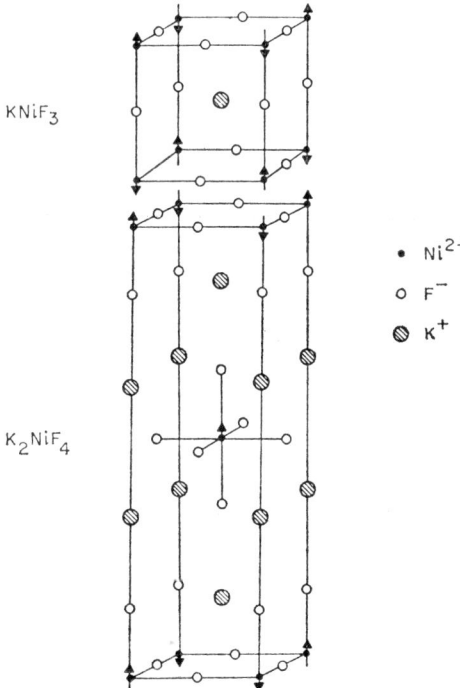

FIGURE 20. The structure and magnetic ordering in $KNiF_3$ and K_2NiF_4 (Birgenau, et al. 1969, 1970).

beginning to look a little two-dimensional. Actually it is not very two-dimensional since the layers are still quite close together and must be coupled magnetically. But there is something special that helps us here. To see this, suppose we have antiferromagnetic ordering within a layer—as indeed one finds—let us ask which way does the central spin in the figure want to point? It looks at the layer above where it sees four equally distant and fully equivalent neighbors; owing to the antiferromagnetic layer ordering, half of these vote "up" and half vote "down". It is just the same with respect to the layer below. Thus the central spin finds itself in the position of the proverbial ass exactly halfway between two equal piles of hay: it does not know which way to point relative to the other layers. So it makes up its own mind within its own layer, completely independently of the other layers. This is just what we would like to happen for two-dimensional behavior. So this is a sort of enforced two-dimensionality which, as I will demonstrate really does occur, as people hoped when they first looked at these structures.

One way we can check this is seen in Figure 21 which shows susceptibility measurements made by Professor Stout in Chicago; the material is MnF_2, a good old three-dimensional antiferromagnet, one of the best! Notice the characteristic shape of the curve for the parallel susceptibility. As T falls it rises sharply and then rounds over, but only within three or four percent of

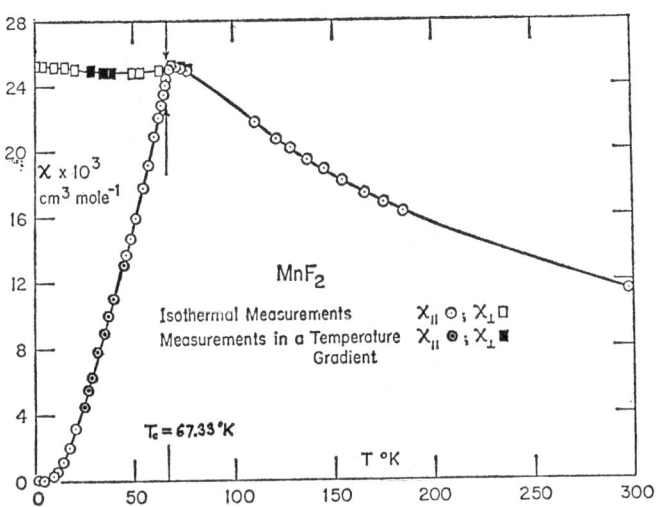

FIGURE 21. The susceptibilities of MnF_2: note the parallel susceptibility (circles). (By courtesy of J. W. Stout, University of Chicago.)

the critical point, which is marked by the arrows; then it drops abruptly. The next picture (Figure 22) is a theoretical one: the curve marked (a) is for a three-dimensional Ising model antiferromagnet and it is quite similar in shape to the curve for MnF_2: it comes up to a fairly sharp maximum near, but 5 or 7% above, T_c and then drops sharply down at T_c. Curve (b) is for a two-dimensional Ising model; notice that it has a broad, flat maximum which is some 40 or 50% above T_c; it slowly keels over and finally drops steeply. The critical point, marked by a small circle, is low down on the curve. So here is a very characteristic quantitative difference depending on dimensionality. Now look at two homologues of K_2NiF_4, namely, Rb_2MnF_4 and K_2MnF_4 which both have the structure shown in Figure 20 but with manganese ions replacing the nickel ions. Figure 23 shows susceptibility measurements made by Breed (1967) in the Netherland. Notice the location of the critical points at 40 to 45°K, and that both curves have very broad, rounded maxima occurring at about 70 to 80°K. The positions of these maxima provide strong evidence that the layers are behaving two-dimensionally, interacting independently of one another.

You might, however, feel that these results are just indicative, but recently at Brookhaven, Birgenau, Guggenheim and Shirane (Birgenau et al., 1969)

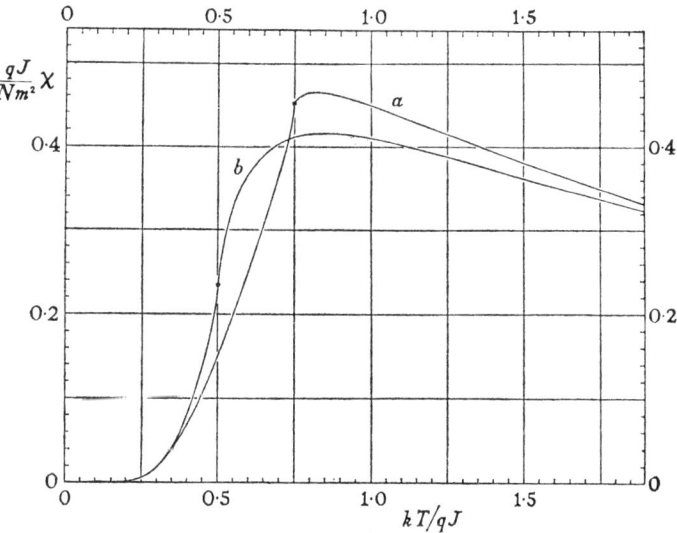

FIGURE 22. The parallel susceptibilities of antiferromagnetic Ising models: (a) simple cubic lattice; (b) plane honeycomb lattice. The critical points are indicated by small circles (Fisher and Sykes, 1962).

have been making a neutron scattering study which has isolated a number of features that really demonstrate conclusively the two-dimensional character of the ordering. They looked at the magnetic Bragg scattering and they also looked above T_c, where one expects to see diffuse scattering because the system is getting ready to order. In a normal antiferromagnet in which the spins are going to order as a three-dimensional array—three-dimensionally packed sardines, up-down-up-down—one sees a diffuse scattering peak, localized at one point in reciprocal space. If you scan across reciprocal space in a suitable direction, then you do see the expected peak as shown in Figure 24; but if you scan in other directions, you find that instead of the diffuse critical scattering being concentrated in a *peak* it is spread out over a diffuse scattering *ridge*. The upper part of Figure 24 shows the growth of the ridge as $T \to T_c+$. The reason for the ridge is simply that along the crystalline z-axis there is no correlation, no significant short range order; the layers of spins are fluctuating quite independently. Here you have very clear and direct evidence that the layers are essentially non-interacting; ordering takes place only within the layers and over a rather broad temperature range.

At the critical temperature of 97.1°K—this is now K_2NiF_4—long-range order finally appears. But at this stage does the crystal still just order in independent two-dimensional layers? Is there no ordering between the layers?

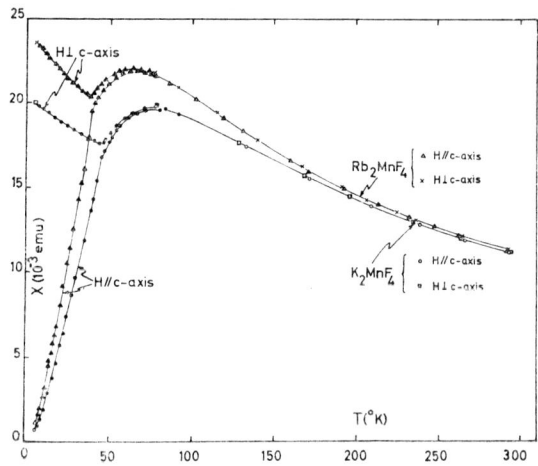

FIGURE 23. The susceptibilities of Rb_2MnF_4 and K_2MnF_4 (Breed 1967).

The answer is that the crystal orders *three*-dimensionally: each layer orders but, in addition, the spins in different layers become fixed with respect to one another. You can see, however, why that is not so surprising by recalling the behavior of the weakly coupled Ising chains. Even when the interchain coupling energy J' was only one-hundredth of the intrachain interaction J, the system ordered two-dimensionally at some relatively high temperature (see Figure 12). Essentially, once the (sub-lattice) spins in a layer have coupled strongly to one another to form an effective macroscopic spin, then even infinitesimal coupling will lock different layers together. Experimental evidence in K_2NiF_4 indicates that the relevant coupling between the layers is down by a factor of 200 or more. Thus, it has a negligible effect on the short-range ordering above T_c, as measured by the diffuse scattering, but it does lock the order in below T_c.

Let us now inquire about the height of the Bragg peak below T_c, which is a direct measure of the long-range order. The observations on the $(1, 0, 0)$ line are illustrated in Figure 25. The height can be measured fairly

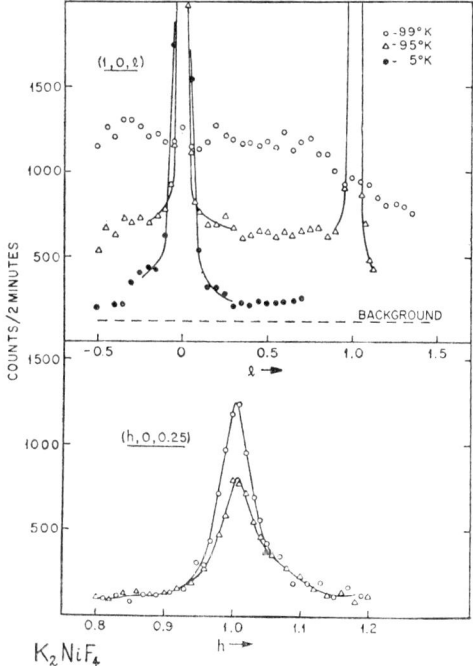

FIGURE 24. The diffuse neutron scattering intensity from K_2NiF_4 above T_c showing the growth of the scattering ridge (Birgenau et al., 1970).

precisely (although this was a preliminary run) and by the arguments I sketched before, its square root should vary as the spontaneous sub-lattice magnetization. Figure 26 shows the behavior on a broader scale all the way down to 0°K, principally to demonstrate that at low temperatures one obtains a good fit to a spin-wave theory. This implies that the system is Heisenberg-like rather than Ising-like. Again, by looking at the spin-wave dispersion relation, which has recently been done, one can check directly that these are two-dimensional spin-waves—they run around in a layer but never jump across—the dispersion relation is independent of the z component of momentum, an impressive demonstration of the reduced dimensionality.

Returning to the region near T_c we ask for the value of the critical exponent β. Figure 27 is a log–log plot of the data: there are three curves for the two-dimensional Ising model which you remember has an exponent $\beta = \frac{1}{8}$; for manganese fluoride which, as we saw before, has $\beta \simeq 0.33$ and, finally, for the potassium nickel fluoride, K_2NiF_4. As you can clearly see, this plot is typical of two dimensions rather than of three! Thus, although the layers are locked together three-dimensionally, their ordering—the way

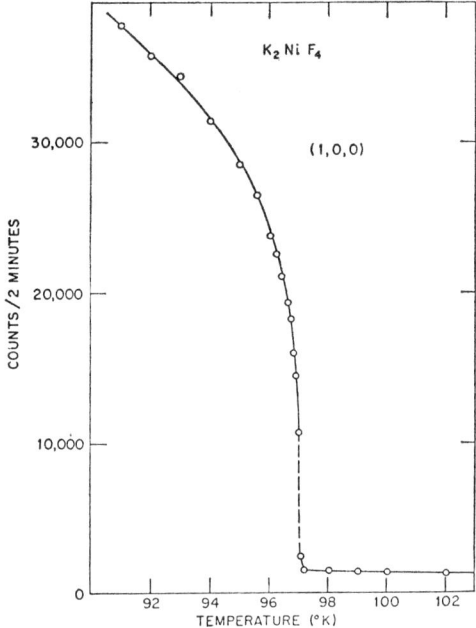

FIGURE 25. Intensity of the Bragg sublattice peak in K_2NiF_4 as a function of temperature near T_c (Birgenau et al., 1970).

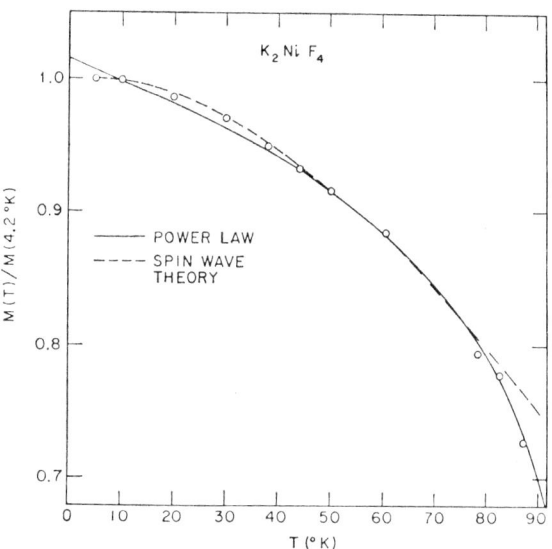

FIGURE 26. The sub-lattice magnetization of K_2NiF_4 showing the fit to spin wave theory at low temperatures (Birgenau et al. 1970).

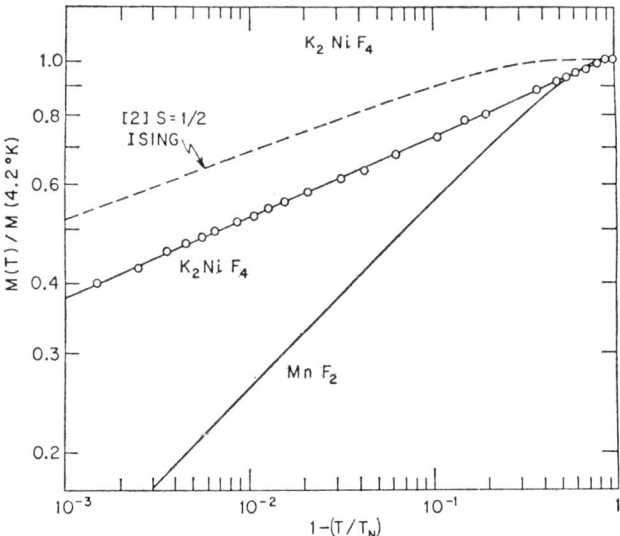

FIGURE 27. Log–log plot of the sub-lattice magnetizations for a two-dimensional Ising antiferromagnet, for K_2NiF_4, and for MnF_2 (Birgenau et al., 1970).

in which the long-range order builds up with temperature—is apparently still characteristically two-dimensional. The observed critical exponent for K_2NiF_4 is not quite $\frac{1}{8}$, as it is in the two-dimensional Ising model. (The fact that we have spin waves proves that the system really cannot correspond very closely to the two-dimensional Ising model.) The value found is $\beta \simeq 0.14$; but these are not really definitive experiments. For Rb_2NiF_4 the observations gave $\beta \simeq 0.16$; the difference between these two values probably gives one some idea of the experimental uncertainties. In any case, the values of β are completely different from those for typical three-dimensional systems like MnF_2, clear evidence of the pseudo-two-dimensionality.

I have pointed out that the early departure of the sub-lattice magnetization from its zero temperature value is a strong indication of the spin-wave, Heisenberg-like character of the interactions in K_2NiF_4. But did I not just state, "If a system was Heisenberg and two-dimensional, it should not have a phase transition"? Indeed, I did claim that. What is the answer to this paradox? The true answer is that we are still busy thinking about what to say! There are various possibilities but probably the correct conclusion is that the spin Hamiltonian is not completely isotropic. What, then, is the ratio of the anisotropy field to the exchange field? It is pretty small: by present estimates, less than one in 500. But, nevertheless, when you are snatching at straws even one five-hundredth of a straw is better than no straw! More seriously, as we learned with the Ising model, since a two-dimensional Heisenberg system is probably on the borderline of wanting to order, just that little bit of anisotropy should be enough to stabilize the spins and give us a real transition that, in a sense, is Heisenberg-like, but in another sense is more Ising-like since there is some anisotropy so that the crystal really has a preferred axis. That may be why the exponent β ($\simeq 0.14$) is not far from the Ising value: it is also possible that if one could go closer towards the critical point and look more carefully, one would see precisely the Ising result $\beta = \frac{1}{8}$. On the other hand, the story *may* be that these "Bragg peaks" are not truly Bragg peaks. Thus, suppose one had the suggested infinite initial susceptibility: that would imply a very slow decay of $\sigma(R)$. On Fourier transformation, to compute the scattering, this means that there would be no delta function peak at the origin but, nevertheless, there would be a very sharp peak, possibly like an inverse power law. Perhaps by going back and looking at the experimental line shape carefully, you could confirm this and even find the actual power law characterizing the divergence of the susceptibility in reciprocal space. The corresponding decay of $\sigma(r)$ would have to satisfy the inequalities (32) but maybe we are already seeing this unusual 'orderless transition'.

These are some of the reasons why these materials are rather exciting and why one is interested in playing with them a lot further, both theoretically and experimentally. The last figure, in fact, illustrates what might be called a "changeover of dimensionality". Displayed in Figure 28 is a log–log plot of spontaneous magnetization with the same scales as in Figure 27 but in this case for yet another structural homologue of K_2NiF_4, namely, rubidium ferrous fluoride, Rb_2FeF_4. In the temperature range from about 30% to 3% below T_c the behavior of the plot, expecially its small slope, is very similar to the other "two-dimensional" materials. But about 2 to 3% below T_c the ferrous fluoride realizes that it really is three-dimensional after all and the slope increases to indicate an exponent β of 0.3 to 0.4. Conversely, if we return to the nickel fluorides, we would hence guess that if one could approach another few decades closer to T_c, we might ultimately see a similar turnoff on the graph to a typically three-dimensional slope asymptotically close to the transition point. Because of the changeover, the ferrous fluoride is more difficult to analyse experimentally; this is seen in the scattering above T_c as well. The diffuse ridges come up and then go down again because the spins really do not want to order only within the layers. Close enough to T_c, ordering occurs across from one layer to

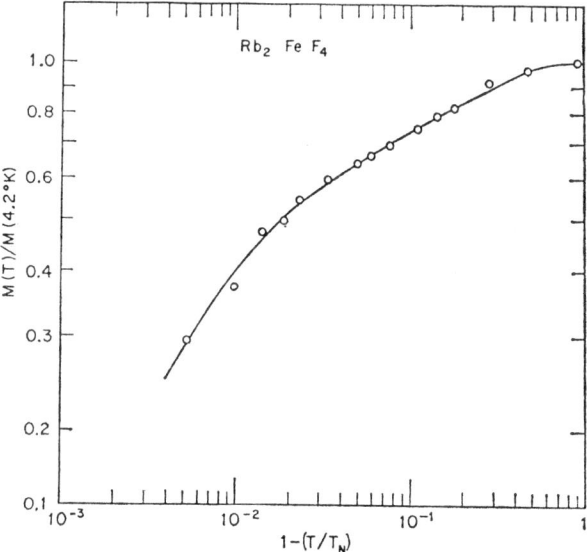

FIGURE 28. The sub-lattice magnetization of Rb_2FeF_4 on the same scales as in Fig. 27 demonstrating the "changeover of dimensionality" (Birgenau *et al.*, 1970).

another so that what was a diffuse scattering ridge turns itself into a diffuse scattering peak. There are a number of other physical systems where you can also see such a "changeover". It would be interesting to try to understand them in more detail and actually correlate the changeover region with the *interlayer* coupling, J'. Experimentally, we might ask if one could alter the 2% figure by, for example, pressure, or by doping?

There is certainly going to be a lot of fun in exploring such behavior further. One of the obvious challenges is to try and make a crystal a bit more two-dimensional by putting more potassium or rubidium in it to force the magnetic layers further apart and thus see how far one can really approach the ideal situation. In this connection, of course, something that the theorists have always wanted to see done, is a study of some *real* two-dimensional systems. Probably one will have to go away from magnetism here. Whenever I say "magnetic thin films", all my friends in magnetism look at me in horror, shudder and say: "Well, they're very interesting from a technological angle, but it's safer not to think about them theoretically." And that tends to be true of many film materials; otherwise one's first candidates for two-dimensional systems would be films. But I think that's going to change as technologies improve. Soon we should be able to look at clean, homogeneous film materials, perhaps measuring specific heat and doing other things with them. In the area of chemistry, films are a little easier to study. One might look harder at adsorbed gas layers on suitable substrates. Again, one tends to get stuck in other nasty problems but there is a lot of interesting work to be done.

In summary we have a number of good ideas concerning the way in which symmetry and dimensionality enter the picture. We have got one or two fully rigorous theorems; but although we are in quite good shape on that count, we do not yet have many rigorous *positive* theorems that really tell us in detail about phase transitions. We do not have any proper solution of the real puzzle as to the values of those mysterious pure numbers—the critical exponents (see the reviews by Kadanoff *et al.*, 1967 and Fisher, 1967). We know they depend on dimensionality, but in what way? Is it simply some inverse power of the dimensionality? What is the formula for the exponent $\beta(d)$ which would give us the right answer when $d = 2$ or 3? And, since I am a theoretician, when $d = 4$, 5, or 6—we will always be able to beat the experimentalists on that one!

XI. Postscript: 1972

In recent months there has been striking progress towards answering the question raised at the end of my essay, namely, the behavior of the critical point exponents as a function of the dimensionality d. In addition, the role of the symmetry enters the theory explicitly through the number n, which counts the number of components of the order parameter: $n = 1$ for a simple, scalar, Ising-like system; $n = 2$ for XY-like systems or superfluids, with cylindrical symmetry; and $n = 3$ for Heisenberg-like systems with isotropic, spherical symmetry. The basic theoretical development is the so-called renormalization group approach to the theory of the critical point advanced last year by my colleague Kenneth G. Wilson (Wilson, 1971). This theory can be combined with the idea of perturbation around four dimensions based upon writing

$$d = 4 - \varepsilon,$$

and treating ε as a small, continuous parameter. Mathematically, this can be done fairly readily even though only integral values of ε have direct physical meaning! One can then obtain formulae for the critical exponents as series in powers of ε. For the exponent γ, which determines the divergence of the susceptibility according to

$$\chi(T) \sim (T - T_c)^{-\gamma}, \qquad T \to T_c+,$$

the first results were

$$\gamma = 1 + \tfrac{1}{6}\varepsilon + \ldots \quad \text{for} \quad n = 1 \quad \text{(Ising-like)},$$
$$= 1 + \tfrac{1}{5}\varepsilon + \ldots \quad \text{for} \quad n = 2. \quad \text{(XY-like)}.$$

(Wilson and Fisher, 1972). In these formulae ε must be positive (or zero) so that $d \leq 4$; for dimensions exceeding 4 (i.e. $\varepsilon < 0$) the exponents stick at the 'classical' values, in this case $\gamma = 1$. Following this, the results have been extended to general n (Fisher and Pfeuty, 1972; Wegner, 1972), and calculated to higher order in ε by Wilson (1972) using special Feynman graph techniques. The latest result is

$$\gamma = 1 + \frac{(n+2)}{2(n+8)}\varepsilon + \frac{(n+2)(n^2 + 22n + 52)}{4(n+8)^3}\varepsilon^2 + O(\varepsilon^3).$$

When this expression is evaluated for $\varepsilon = 1$, corresponding to three-dimensions, one finds $\gamma \simeq 1.244$ for Ising-like ($n = 1$) systems; the best numerical estimates, based on high temperature series expansions (see e.g. the review, Fisher, 1967), yield $\gamma = 1.250 \pm 0.003$. The agreement is remarkable! For Heisenberg-like systems ($n = 3$) the ε-expansion yields $\gamma \simeq 1.347$, whereas numerical estimates give $\gamma = 1.38 \pm 0.02$; here the higher order terms in ε presumably play a slightly larger role. An especially encouraging feature of these new developments is that the theory gives the correct qualitative form of dependence *both* on the dimensionality d, and on the symmetry number n.

Acknowledgements

The new theoretical research work described in this essay was supported by the National Science Foundation and by the Advanced Research Projects Agency through the Materials Science Center at Cornell University. The manuscript was prepared during the tenure of a John Simon Guggenheim Memorial Foundation Fellowship as a visiting professor in the Department of Applied Physics at Stanford University (partly supported by the Army Research Office, Durham, North Carolina, U.S.A.). The author is grateful for the support granted, and for the hospitality of Professor S. Doniach at Stanford.

Bibliography

Fisher, M. E. (1967). *Rep. Prog. Phys.* **30**, 615.
Kadanoff, L. P., Götze, W., Hamblen, D., Hecht, R., Lewis, E. A. S., Palciauskas, V. V., Rayl, M., Swift, J., Aspnes, D. and Kane, J. (1967). *Rev. Mod. Phys.* **39**, 395.
Lieb, E. H. and Mattis, D. C. (1966). Mathematical Physics in One Dimension, Academic Press, New York and London.
Ruelle, D. (1969). "Statistical Mechanics: Rigorous Results", W. A. Benjamin Inc., New York.

References

Ahlers, G. (1971). *Phys. Rev. A* **3**, 696.
Birgenau, R. J., Guggenheim, H. J. and Shirane, G. (1969a). *Phys. Rev. Lett.* **22**, 720.
Birgenau, R. J., Guggenheim, H. J. and Shirane, G. (1969b). *Bull. Am. Phys. Soc.* **6**, 738.
Birgenau, R. J., Guggenheim, H. J. and Shirane, G. (1970). *Phys. Rev. B* **1**, 2211.
Bogoliubov, N. N. (1962). *Physik Abh. Sovj. Un.* **6**, 1; **6**, 113; **6**, 229.
Breed, D. J. (1967). *Physica* **37**, 35, and Ph.D. Thesis (Leiden).
Chester, G. V., Fisher, M. E. and Mermin, N. D. (1969). *Phys. Rev.* **185**, 760.
Dobrushin, R. L. (1965). *Dokl. Akad. Nauk. SSSR.* **160**, 1046 (English Translation Soviet Phys. Dokl. **10**, 111)
Domb, C. (1960). *Adv. Phys.* **9**, 150.
Fairbank, W. M., Buckingham, M. J. and Kellers, C. F. (1957). *Proc. Fifth Int. Conf. on Low Temp. Phys.* p. 50., Madison, Wisconsin.
Ferdinand, A. E. and Fisher, M. E. (1969). *Phys. Rev.* **185**, 832.
Fisher, M. E. (1964). *Arch. Ration. Mech. Anal.* **17**, 377.
Fisher, M. E., and Pfeuty, P. (1972), *Phys. Rev. B*. (In press).
Fisher, M. E. and Sykes, M. F. (1962). *Physica* **28**, 939.

Griffiths, R. B. (1964). *Phys. Rev.* **136A**, 437.
Haseda, T. and Miedema, A. R. (1961). *Physica* **27**, 1102.
Heller, P. and Benedek, G. B. (1962). *Phys. Rev. Lett.* **8**, 428.
Hohenberg, P. C. (1967). *Phys. Rev.* **158**, 383.
Jasnow, D. and Fisher, M. E. (1971). *Phys. Rev. B.* **3**, 895, 907.
Mermin, N. D. and Wagner, H. (1966). *Phys. Rev. Lett.* **17**, 1133.
Onsager, L. (1944). *Phys. Rev.* **65**, 117.
Peierls, R. E. (1936). *Proc. Camb. Phil. Soc.,* **32**, 477.
Rushbrooke, G. S. and Wood, P. J. (1955). *Proc. Phys. Soc.* **A68**, 1161.
Rushbrooke, G. S. and Wood, P. J. (1958). *Mol. Phys.* **1**, 257.
Stanley, H. E. and Kaplan, T. A. (1966). *Phys. Rev. Lett.* **17**, 913.
Stevens, C. L. and Felsenfeld, G. (1964). *Biopolymers* **2**, 293.
Wegner, F. (1972). *Phys. Rev. B.* (In press).
Wilson, K. G. (1971). *Phys. Rev. B* **4**, 3174, 3184.
Wilson, K. G. (1972). *Phys. Rev. Lett.* **28**, 548.
Wilson, K. G. and Fisher, M. E. (1972). *Phys. Rev. Lett.* **28**, 240.
Yuval, G. and Anderson, P. W. (1969). *Phys. Rev. B* **1**, 1522.

Arc Physics

A. M. HOWATSON

University of Oxford, Engineering Science Department, Parks Road, Oxford

I. Introduction	92
A. Historical	92
B. The arc defined	92
C. Establishing an arc	93
D. Elementary arc structure	94
II. Thermal equilibrium in the arc	95
A. The temperature difference between electrons and gas	96
B. The approach to equilibrium distributions	97
C. Populations of species	98
D. The effects of optical thinness and diffusion	100
E. Summary	102
III. Single-fluid models of the high-pressure column	102
A. The wall-stabilized arc	104
B. Electrode-stabilized arcs	105
C. Convection-stabilized arcs	106
D. Free-burning arcs	108
E. The channel model	109
F. Plasma jets	109
G. Gas properties	109
H. The dynamic behaviour of the column	112
IV. The electrode regions	113
A. The cathode region	114
B. The anode region	117
V. Arc measurements	117
A. Electrode measurements	117
B. Column measurements	118
References	122

I. Introduction

A. HISTORICAL

The electric arc has been well-known for a century and a half; for the greater part of that time it has played an important role in several practical devices. It first received extensive scientific scrutiny some seventy years ago, with experiments by Mrs. Ayrton (1902) on the carbon arc, and ever since has been the subject of determined investigation. Early measurements were confined mainly to current–voltage characteristics, but by the 1930's it was feasible to measure the temperature and density of the arc plasma, and experimental techniques have since improved steadily; at the same time, theoretical models have been developed. Nevertheless, the arc is a complicated phenomenon and in consequence, while it has features which are now well-understood, there are others which are well-known but understood hardly at all. The most significant development in recent years has been an improved understanding of the role of gas flow (which, owing to natural convection or otherwise, is rarely absent) in the heat transfer from the arc column.

B. THE ARC DEFINED

The arc may best be distinguished from other self-sustaining discharges by its low voltage. Although under certain conditions its voltage may reach many hundreds of volts, it is more usually measured in tens; this is due to the low cathode fall of potential, of order 10 V, and this in turn is connected with the nature of electron emission from the cathode. A glow discharge has a quite different cathode mechanism and a cathode fall of order 100 V or more.

It is also possible to define an arc on the basis of thermal equilibrium (in the sense that gas and electron temperatures are equal) in the positive column. This is possible only at high pressures (above about 0.1 bar) and therefore excludes the low-pressure arc, which shows the same low cathode fall, while it includes the high-pressure glow with its different cathode behaviour. However, the high-pressure glow is a relatively rare and unstable discharge, and the column of a low-pressure arc is little different from that of a glow discharge at the same pressure; so it is true to say that interest in arcs relates mostly to those high pressures at which thermal equilibrium is approached (but not infallibly attained). In this respect, arc physics becomes radically different from other aspects of plasma physics.

The distinction between arc and glow discharges in terms of cathode behaviour disappears at frequencies high enough for the importance of the electrodes to diminish: roughly, when the period of a cycle becomes comparable with the transit time of a charge carrier, or upwards of 1 MHz.

The low voltage which is characteristic of an arc can be readily achieved with an externally-heated thermionic cathode in a gas-filled tube (for example, the thyratron valve). This is appropriate only to low pressures, and is not truly a self-sustaining discharge.

In what follows we deal with the self-sustaining arc, at high pressure and at d.c. or low frequencies, but do not necessarily exclude transient behaviour.

C. ESTABLISHING AN ARC

An arc may be established by drawing apart current-carrying electrodes which are in contact, providing that the circuit voltage is sufficient to maintain an arc; or by applying to two electrodes a voltage sufficient to break down the gap between them, forming a spark which becomes in the steady state an arc; or by effecting a transition from an existing glow discharge. The glow-to-arc transition may be brought about either by increasing current at constant pressure, or by increasing pressure at constant current.

The transition to an arc is favoured by an increase in pressure because both current density and voltage gradient in the cathode region are increased in consequence, and so therefore is the local heating. These conditions favour one or other of the arc cathode mechanisms. The state at which a transition will occur for a given current depends on the cathode condition; the transition can be inhibited by cathode cooling and by using very pure, clean electrodes.

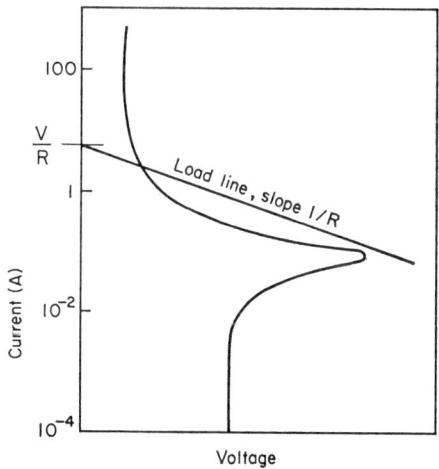

FIGURE 1. The voltage–current characteristic of a static discharge.

The transition brought about by increasing current can be traced on the characteristic shown in Fig. 1, which is typical for a d.c. discharge at low to medium pressure. The operating point on the curve is evidently determined by the voltage V and impedance R of the supply circuit, represented by the load line. The peak of voltage which represents the transition can be overcome, to achieve an arc, only if the source voltage is high enough and its impedance low enough; the form of the curve is such that even then the transition, once past the peak, may take the form of an uncontrolled jump in current to a stable arc. Here again, the approach to transition is a region of increasing current density and voltage gradient (the so-called abnormal glow region) which results in cathode heating; so the precise form of the transition peak is affected by the purity, surface condition and degree of cooling of the cathode, as well as by its material.

In theories of the arc, it is usually convenient to adopt current as an independent variable, except in certain transient situations.

D. ELEMENTARY ARC STRUCTURE

The approximate form of voltage distribution along the length of a typical arc is shown in Fig. 2. There are three main regions.

Adjacent to the cathode is the cathode fall of potential, of order 10 V, which occurs over a very short distance (in the order of 10^{-1} mm or less) and embraces a small region of positive net space charge and very high current density (10 to 10^4 A/mm^2, or more) in the form of one or more cathode spots. This region is crucial to the electron emission from the cathode which maintains the discharge; and it can be regarded as a solid-to-gas junction having something in common with other forms of current-carrying junction.

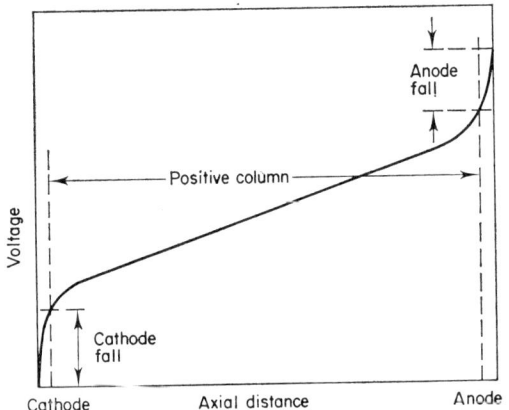

FIGURE 2. The axial distribution of potential in an arc.

Adjacent to the anode, similarly, is the anode fall of potential, also embracing a very small region of net space charge in one or more spots of high current density; but here the space charge is negative and there is no anode emission.* The anode fall and current density are normally rather less than those of the cathode.

Between these two electrode regions is the arc column, traditionally called the positive column, a region of plasma which is electrically neutral or nearly so, with a degree of ionization which in the hottest parts may be anything from less than 1% to more than 50%. In the column the axial electric field is more or less uniform, and usually rather higher than in a typical glow discharge. Figure 2 shows longer regions of non-uniform field than are required to represent the electrode falls; for the column must include at its ends transition regions, where it not only converges in area to the electrode spots but may also be appreciably cooled by the electrodes. The electrodes are hot, but in a high-pressure arc the gas in the middle of the column is much hotter: temperatures of 10,000 K are common. Nor should we assume that an apparently uniform axial field means that the column in that region is necessarily uniform. A truly uniform column, though often postulated, is in practice the exception rather than the rule; but the variation of the axial field in the greater part of it is not often sufficient to exceed uncertainties in measurement.

We deal below with these several regions in more detail.

II. Thermal Equilibrium in the Arc

The electric field in a gas discharge imparts kinetic energy to charged particles. This energy in turn is transferred to neutral particles by collisions, which heat the neutral gas to a greater or lesser degree and tend to randomize the higher energies of the charge carriers. Being much lighter than ions, electrons not only receive more energy in a given interval but also (by purely classical mechanics) can transfer a smaller fraction of their total energy in an average collision. As a result, at low pressures the electrons have much higher average energies than other particles, sufficiently random (unless at very low pressures) to represent a high temperature, while the other particles are heated only to a small degree. In a low-pressure column electron temperatures are typically more than 10^4 K while ion and neutral gas temperatures are scarcely above ambient. If the plasma is strongly ionized, collisions with neutrals no longer dominate all others but the same result holds.

---

* An exception to this is the Beck arc, which has an anode designed to emit positive ions.

At higher pressures, mean free paths are reduced and collision frequencies increased. Electrons gain less energy between collisions but transfer a certain fraction of their energy with greater frequency. In consequence the temperature difference between electrons and the remaining gas is less, and the gas is hotter. At pressures of around 0.1 bar and above the temperature difference between the two may become small enough to ignore, and we have only a single plasma temperature at any point. Figure 3 illustrates this.

There are then two basic aspects to thermal equilibrium: one, that the electron and gas temperatures are equal; the other, that the energy distributions which depend on temperature (and on which many deductions rest) are in fact those of statistical equilibrium (that is, that they are in general of Maxwell–Boltzmann form: except for photons, conditions in a gaseous plasma are such that classical statistics suffice).

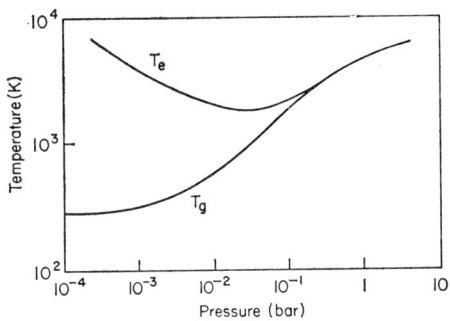

FIGURE 3. The convergence of electron and gas temperatures.

A. THE TEMPERATURE DIFFERENCE BETWEEN ELECTRONS AND GAS

The difference between the electron and gas temperatures T_e and T_g may be assessed by an argument based on the rate of energy transfer from electrons to gas. The electrons have an average energy which exceeds that of the gas particles by $\tfrac{3}{2}k(T_e - T_g)$; if they lose an average fraction δ of this at a collision and if their collision frequency is v, then the rate of energy transfer per electron is $\tfrac{3}{2}\delta v k(T_e - T_g)$. This may be equated to the electrical power input per electron, which for an electric field E can be written $E^2 e^2/mv$ if we recall that the mobility is usually written as e/mv. So we arrive at

$$T_e - T_g = \frac{2}{3} \frac{E^2 e^2}{\delta k m v^2}, \tag{1}$$

which shows immediately the strong effect of the field in raising the temperature difference and of the collision frequency (i.e. of pressure) in reducing it. The expression may be cast in a number of other forms: for example, it is common to write δ as $2m/M$ for electrons of mass m in gas particles of mass M. This is the average energy fraction transferred by electrons in elastic collisions with a classical stationary gas; in a gas discharge it is in error to the extent to which inelastic collisions are important (there is also a slight error due to the non-isotropic nature of elastic scattering). Although inelastic collisions are almost always in the minority, they may dominate energy transfer when $T_e - T_g$ is large (Wasserab, 1950); but in the high-pressure column the temperature difference is usually such that their contribution is small, and we may with sufficient accuracy set δ to $2m/M$. If we also express ν as the ratio of mean velocity to mean free path λ (for the electrons), taking the former to be $\sqrt{(8kT_e/\pi m)}$ as for a Maxwell distribution, then we get finally

$$\frac{T_e - T_g}{T_e} = \frac{\pi}{24}\frac{M}{m}\left(\frac{Ee\lambda}{kT_e}\right)^2. \qquad (2)$$

The difficulty in using eqn (2) lies in deciding values for λ and T_e; both can be roughly estimated, but a reliable value for λ needs a knowledge of elastic cross-sections at the appropriate electron energies. The temperature difference according to the above expressions is by no means negligible, even at atmospheric pressure, in arcs with high voltage gradients of order 10 kV/m.

B. THE APPROACH TO EQUILIBRIUM DISTRIBUTIONS

Perfect equilibrium can be achieved only in a black-body cavity; in any practical arc device there exists an inflow and an outflow of energy which result in gradients of electric field and of temperature; the randomness of equilibrium is disturbed by the directed flows of particles and of energy which result. One can assert that the disturbance is negligible if the change in any quantity over distances in the order of λ is small; thus we may require that, with any distance coordinate r,

$$\lambda \frac{\partial T}{\partial r} \ll T,$$

and

$$\lambda \frac{\partial n}{\partial r} \ll n,$$

for any species having temperature T and number density n, and that

$$Ee\lambda \ll kT$$

for any charged species. These conditions restrict heat transport, diffusion flux (or mass transport) and electrical energy transfer (or charge transport) respectively. If they are fulfilled, local thermal equilibrium (LTE) may be achieved, so far as the effect of gradients in the stationary arc is concerned. There is, of course, a comparable restriction on rapid time variation.

It is possible to calculate modified distribution functions, for example that which results from the field E acting on electrons which suffer certain kinds of collision (see, for example, Druyvesteyn and Penning, 1940). In low-pressure discharges the distribution of electron energies can be measured by Langmuir probes, and frequently shows considerable departures from the Maxwell form. However, in arcs we are usually less concerned with the precise form of the distortion of a velocity distribution function than with the effect of non-equilibrium on the populations of species; this is an important question for many arc measurements, and we return to it below.

In one important respect, that of radiation, nearly all laboratory plasmas are substantially out of equilibrium; by common consent this is not in itself deemed to be a departure from LTE. Photons in equilibrium obey Einstein–Bose statistics, and yield black-body radiation. In an arc, this state can be approached only at very high pressures, of order 100 bar; at atmospheric pressure, any arc shows strong line emission with a relatively weak continuum far below the black-body level for the temperature of the plasma. Fortunately, the effect of this on the equilibrium of other species need not be serious in a collision-dominated plasma, and line emission can be a valuable aid to measurement (Section V, B). It is important to know the degree of opacity of the plasma. Many measurement techniques rely on an optically thin plasma, with negligible self-absorption; on the other hand, optical thickness with a short absorption length for photons, comparable to a mean free path, would ultimately tend to produce complete local equilibrium and black-body radiation from the plasma surface. The degree of opacity increases with pressure, but is in any case strongly dependent on the wavelength considered.

A general requirement for LTE in an arc is that of sufficient current to prevent the column voltage gradient from becoming too high and the electron density too low. In a typical arc at about atmospheric pressure there is a minimum current, of order 1 A, below which LTE cannot be expected.

C. POPULATIONS OF SPECIES

A consequence of local thermal equilibrium is that the population of all species at any point of known temperature can be found from statistical principles. Thus, if n_{jk} is the number density of a species which dissociates into constituents j and k of number densities n_j, n_k with a dissociation energy ε_{jk}, then the three populations are related by the law of mass action according to the Saha equation,

$$\frac{n_j n_k}{n_{jk}} = \frac{b_j b_k}{b_{jk}} \left\{ \frac{2\pi m_j m_k kT}{h^2 (m_j + m_k)} \right\}^{3/2} \exp\left(-\varepsilon_{jk}/kT\right), \tag{3}$$

in which b represents the partition function (a function of the Boltzmann factors for all energy levels of the species concerned). The equation can be used to estimate electron density by letting j represent electrons, k positive ions, jk neutral atoms and ε_{jk} the ionization potential; in this case $b_j = 2$ while b_k and b_{jk} must be summed over all excited states of the ion and the atom. In weakly ionized monatomic gases one may for rough estimates set b_k and b_{jk} equal to the statistical weights of the ground states, so ignoring excitation. For more complicated situations, as in molecular or chemically reacting gases, the only proper way to calculate populations for N different species is to form a set of $N-2$ equations like (3), for each of which ε_{jk} must be known, together with equations for the conservation of charge or mass and total pressure. The solution of these gives a set of population curves, of which Fig. 4 shows an example for nitrogen. In practice even simple gas systems imply a large number of possible species, especially at high temperatures where multiple ionization is significant. The electron density saturates and (at constant pressure) even declines with increasing temperature as the region of full ionization is approached; other species show clear maxima at certain temperatures, a fact which is responsible for layers of enhanced brightness in the outer regions of some arc columns.

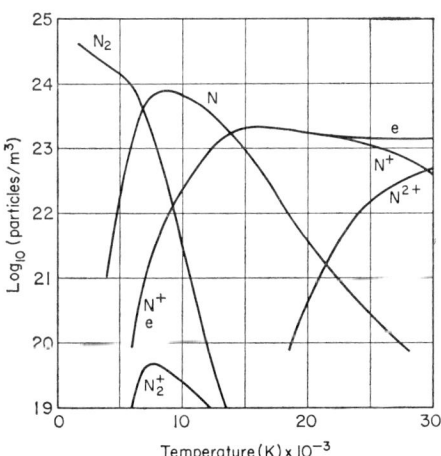

FIGURE 4. The equilibrium composition of nitrogen at atmospheric pressure (after Burhorn, 1959).

A simpler relation between populations is afforded by the Boltzmann law which, for equilibrium, gives

$$\frac{n_j}{n_k} = \frac{g_j}{g_k} \exp\left[-(\varepsilon_j - \varepsilon_k)/kT\right], \tag{4}$$

in which n_j, n_k are number densities of particles in energy levels $\varepsilon_j, \varepsilon_k$ with statistical weights g_j, g_k.

D. THE EFFECTS OF OPTICAL THINNESS AND DIFFUSION

The above relations enable any population to be calculated, at least in principle, for a state of LTE at given pressure and temperature, without any reference to the detailed processes which occur. But they do not allow any estimate of the change in populations caused by any departure from LTE; for this we must resort to another method of calculating a population, namely the solution of rate equations. The rates of population and depopulation of a particular kind of excited particle, say, can be expressed in terms of transition probabilities (for transitions which absorb or emit photons) and of rate coefficients, provided that we deal with a homogeneous plasma so that diffusion does not contribute. For example, the rate of decrease of the number density n_j of atoms in an excitation level j due to colliding electrons (density n_e) producing excitation to a higher level k is given by

$$\frac{\partial n_j}{\partial t} = n_e n_j K_{jk}, \tag{5}$$

in which K_{jk} is the rate coefficient for the process and is given by

$$K_{jk} = \langle \sigma_{jk} v \rangle.$$

Here σ_{jk}, the cross-section for the process, is a function of the electron velocity v. The average is taken by integrating $\sigma_{jk} v$ over the electron velocity distribution, which is with sufficient accuracy assumed to be Maxwellian; thus the electron temperature T_e becomes, as it must, a parameter determining the populations in this case.

We need, then, to identify all important species and the processes which add to and subtract from their numbers; setting all $\partial n_j/\partial t$ to zero for a stationary state yields population ratios as functions of temperature. Many such calculations have been made in recent years; most of them are for homogeneous hydrogen-like plasmas (the complexity grows rapidly when less simple atoms are considered) in which only radiation and electron–atom collisions are deemed to govern the populations, although Drawin (1969) and others have also considered atom–atom collisions. The calculations have shown departures from Boltzmann populations caused by various degrees of optical transparency. Figure 5 shows an example; Richter (1971) gives a review and bibliography.

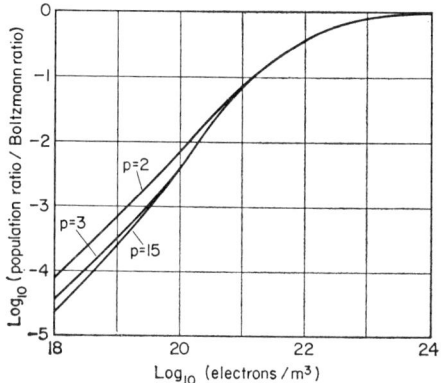

FIGURE 5. The underpopulation of excited levels in optically thin hydrogen at 16000K. The ordinate gives the ratio of numbers in level p to numbers in the ground state, as a fraction of the Boltzmann ratio (after McWhirter and Hearn, 1963).

The results demonstrate that optical transparency tends to under-populate excited states and over-populate the ground state, and that the effect becomes insignificant if n_e is large. It is not difficult to appreciate this: for the simple situation in which only the transition from the ground state 1 to an excited level 2 is important, the rate equation gives a population ratio

$$\frac{n_2}{n_1} = \frac{n_e \langle \sigma_{12} v \rangle}{A_{21} + n_e \langle \sigma_{21} v \rangle}, \qquad (6)$$

for no self-absorption; here A_{21} is the transition probability for spontaneous emission. LTE would require $A_{21} \ll n_e \langle \sigma_{21} v \rangle$ for the ratio then clearly satisfies the detailed balancing principle, which in turn requires the ratio of cross-sections to agree with Boltzmann's law. So a necessary condition for n_1 and n_2 to be within 10% of their equilibrium values is

$$n_e \gtrsim 10 A_{21} / \langle \sigma_{21} v \rangle. \qquad (7)$$

By using approximate relations for σ_{21} and A_{21}, eqn (7) may be reduced to

$$n_e \gtrsim 10^{18} \varepsilon_{12}^3 \sqrt{T_e} \, \mathrm{m}^{-3}, \qquad (8)$$

where ε_{12} is the transition energy in eV (McWhirter, 1965).

The foregoing arguments enable departures from LTE to be assessed for a homogeneous plasma. But no laboratory plasma is homogeneous; diffusion inevitably occurs and should properly appear in the rate equations in the form of terms like $\nabla . n\mathbf{v}_d$ where $\mathbf{v}_d$ is the drift velocity of diffusion. Now, however, we can no longer rely on a set of local equations, for we are involved in the spatial variation of n and $\mathbf{v}_d$; so we are obliged to consider an overall model for the column, such as those discussed in later sections, as well as detailed atomic processes. Uhlenbusch et al. (1970) have carried out such calculations for wall-stabilized arcs (see Section III, A) in noble gases at pressures near atmospheric. Comparing their results with experimental measurement of line intensities, they concluded that there are important departures from LTE which can be attributed to diffusion. In helium not only is there a significant difference between T_e and T_g, but also the electron density may be far above its equilibrium value in the cooler outer regions of the arc.

E. SUMMARY

It may be clear from the foregoing sections that the question of thermal equilibrium in an arc plasma is not easy; but its crucial inportance justifies a close attention which it does not always receive. It is common practice to assume LTE in all theoretical models of the arc column (to be discussed below) and in spectroscopic techniques of temperature measurement. Were the purpose of these merely to determine the overall behaviour of the arc column, the errors which accrue would probably not be serious; but in fact as we shall see the derivation of basic data for high-temperature gases, e.g. transport coefficients, depends on the measurement of arc temperatures and on their substitution in a reliable theory. So departures from LTE can be more serious than at first sight appears, and their investigation is one of the areas of attack in present-day arc physics. Having said this, we shall from this point deal with theories of the column in which equilibrium is assumed; in many cases, such as those involving gas flow, any other assumption is for the present prohibitive.

III. Single-fluid Models of the High-pressure Column

The assumption of local thermal equilibrium makes it possible to treat the arc column as a fluid in which heat input and heat transport can be expressed by the usual equations of thermodynamics and fluid mechanics. The heat input per unit volume due to an electric fluid $\mathbf{E}$ driving a current density $\mathbf{J}$ at any point is most easily written as $\mathbf{E} \cdot \mathbf{J}$ or $E^2\sigma$ if we suppose that $\mathbf{J}$ is given by $\sigma\mathbf{E}$, where σ is the electrical conductivity at the point in question. It is usual, and often safe, to assume this simplest ohmic relation, but it neglects a number of effects: even in the stationary state, the forces acting on particles include not only the Lorentz force $e(\mathbf{E} + \mathbf{v} \times \mathbf{B})$ but also gravity and diffusion, both thermal and pressure-gradient. Consider, for example, radial diffusion due to the changing concentrations of electrons and ions from the axis of the column outwards, as the temperature falls from some high value to ambient. There can be no net radial current to the non-conducting surroundings of the column; the diffusion is therefore ambipolar and a radial electric field is set up in consequence, which slows the electrons and speeds the positive ions to balance their differing diffusion rates. Further, to conserve mass there must be an inward flow of neutral particles, and the ambipolar diffusion must set up a finite, if small, net space charge. We ignore these effects together with thermal diffusion, gravity (so far as it affects Ohm's law) and induced e.m.f.; and we regard the electric field as simply $\mathbf{J}/\sigma$, which is mainly (but not entirely) axial.

If fluid flow is to be considered, such as results from a pressure gradient or from the buoyancy force of gravity, equations of force or momentum and of continuity must be set up. There is, unfortunately, only a limited class of high-pressure arcs in which gas flow can be ignored. The simplest discharge of all to obtain experimentally, that is the free-burning atmospheric arc, is also the most difficult to represent in theory because of convective flow.

Finally the most important equation of all, which is essential to any model of the arc column, is that of energy balance. It equates the electrical input energy to the energy transported to the surroundings, and in general requires terms for conduction, radiation and convection; it also requires knowledge of transport coefficients as functions of gas properties.

Detailed reviews of the general equations for an arc plasma have been given by Edels (1961) and Maecker (1964). We proceed below to consider their application to particular models representing various kinds of arc. The nature of heat transport from the column and the form of the solution to its governing equations are controlled by its boundaries. In theory an unbounded column expands indefinitely. That it does not do so in practice is due not only to the effect of electrodes but also to the fact that boundaries arise naturally: if no solid boundary, such as a tube wall, is provided then the column is restricted by a flow of cold gas (or even liquid) whether forced or naturally induced. The word *stabilization* is used of an arc not simply to mean the elimination of unsteady behaviour, but also the provision of some boundary of whatever nature to act as a heat sink and to constrain the column. Different methods of stabilization suggest different models and we consider them in turn.

A. THE WALL–STABILIZED ARC

In the wall-stabilized arc a cold boundary is provided at a radius small enough to make radial conduction the chief mode of heat transport. In its experimental form, first developed by Maecker (1956), it consists of a cascade of water-cooled copper annuli, insulated from each other to avoid a short-circuit and forming a continuous tube. The inside diameter is some 10 mm or less and its length of order 100 mm; electrodes are provided at each end. Not only does this cascade provide an intense, stable arc column but it can also be well represented by a simple model and hence used to derive gas properties from temperature measurements. If we suppose convection to be negligible the energy equation of the single-fluid model is

$$\sigma E^2 + \nabla . \kappa \nabla T - P = 0, \tag{9}$$

in which P represents the net loss per unit volume due to radiation and κ the thermal conductivity. σ, κ and P are functions of temperature T. A useful transformation is obtained by introducing the heat-flux potential

$$\phi = \int_0^T \kappa \, dT.$$

If in addition we invoke cylindrical symmetry and neglect radiation as an energy loss, then

$$\sigma E^2 + \frac{1}{r} \frac{d}{dr}\left(r \frac{d\phi}{dr}\right) = 0, \tag{10}$$

in which now σ is a function of ϕ; this is the Elenbaas–Heller equation. If linearized forms for $\sigma(\phi)$ are assumed, closed solutions are possible. For example if $\sigma = 0$ for $\phi < \phi_c$, then $\phi(r)$ is logarithmic for $\phi < \phi_c$; that is, beyond some radius r_c at which $\phi = \phi_c$ and which bounds the electrically conducting column. Within the conducting region $(r < r_c)$ one finds a parabolic $\phi(r)$ for constant σ, or a Bessel form for $\phi(r)$ with linear $\sigma(\phi)$. However, these analytical solutions are of no great value, and it is preferable to compute solutions using numerical data for $\sigma(T)$ and $\kappa(T)$ to obtain $T(r)$. Uniformity in the axial coordinate z requires that $\mathbf{J}$ and $\mathbf{E}$ (since we take $\mathbf{E}\sigma = \mathbf{J}$) are purely axial and constant in z; it then follows that E is constant in r (since $\nabla \times \mathbf{E} = 0$). So having obtained $T(r)$ for chosen E, the arc current I we find as $2\pi E \int_0^R \sigma r \, dr$ where R is the wall radius.

Equation (10) has interesting properties. It can be shown, for example, that the axis temperature T_0 determines, for a given gas, the power gradient EI and the product ER; and a number of relationships in non-dimensional form can be deduced which may also include a radiation loss term (Lord, 1967).

From a measured temperature profile, and knowledge of E and I, one may deduce information on the transport coefficients σ and κ. The conductance G per unit length of column is I/E and can be written

$$G = 2\pi \int_0^R \sigma r \, dr.$$

If now the conductivity σ for a partially ionized plasma is expressed in terms of temperature and an average cross-section (Maecker, 1964) then integration yields a value for the latter with G known and so fixes $\sigma(T)$ and $\sigma(r)$. Equation (9) then gives

$$\kappa(r) = \frac{2\pi E^2}{r|dT/dr|} \int_0^r \sigma(r) r \, dr, \tag{11}$$

from which $\kappa(T)$ follows. We discuss transport coefficients further in Section III, G.

An experimental variant of the wall-stabilized arc is a form in which vorticity in the column is used to eject cold gas and confine hot, so constraining the column and dominating the effect of natural convection. There are several versions: the vortex flow may be induced by tangential gas injection or by rotating the containing tube, and many early measurements were made on such columns. Gerdien (Gerdien and Lotz, 1923) ran an arc in a vortex of water, and subsequent measurements on Gerdien arcs have yielded the highest stationary temperatures achieved in the laboratory, around 50,000 K (Burhorn et al., 1951), by the use of tubes a few millimetres in diameter and currents above 1 kA. (These conditions would be too severe for a normal cascade arc, although careful design of water-cooled copper rings can now achieve heat conduction in excess of the traditional limit of 10 kW/cm^2.) Circulating flow does not invariably result in enhanced temperatures: an arc chamber several centimetres in diameter with a rotating wall gives an expanded and relatively cool column because the flow is induced near the wall.

B. ELECTRODE–STABILIZED ARCS

If the arc column is short, of the same order as the electrode diameter, axial uniformity is no longer possible because of heat conduction to the electrodes. If the gap is short enough for this to be significant, then convection may be negligible even in the free-burning case, and eqn (9) applies. Rompe et al. (1944) solved this case in elliptical coordinates, showing that the thermal field gave spherical symmetry at large distances.

C. CONVECTION–STABILIZED ARCS

We now come to a large class of arcs in which there is appreciable gas flow. Some importance has been attached to arcs in transverse flow, especially the case of the magnetically-driven arc in still air, but the axial-flow situation has predominated in theory and in experiment. (It is worth noting, however, that transverse natural convection gave the arc its name.) Axial flow includes the natural convection of a vertical free-burning arc as well as forced convection, in a tube or nozzle or otherwise. The boundary conditions for the column are formed not by a fixed-temperature wall but by a free stream of cold gas. This leads us to the most complete single-fluid model of the arc, requiring additional terms in the energy eqn (9), of the form $\rho c_p \mathbf{v} . \nabla T$ where ρ, c_p and $\mathbf{v}$ are the mass density, specific heat at constant pressure and local macroscopic velocity of the arc fluid. Solutions for this form are more difficult and require further approximations compared with the wall-stabilized case. The conduction term must be retained, for even in forced convection it remains significant at least in the hot central region; indeed several solutions postulate that in the current-carrying core convection is always negligible, while the complete equation applies to the outer region of the column.

The over-riding effect introduced by convection in the energy equation for axial flow is the loss of axial uniformity. It is clear that convection must cause the isotherms to expand with distance z in the flow direction as the cold gas is heated by the column; and the scalar product in the convection term shows that it can exist only where there is a component of gas flow in the direction of ∇T. The use of the term 'fully-developed', sometimes used of any uniform arc column, now becomes significant: in axial flow, the fully-developed column is one which is constricted and long enough for the isotherms and the gas velocity vector to be everywhere axial, so that convective transport falls to zero and the input power gradient EI is balanced by conduction and radiation only; there is therefore axial uniformity and eqn (9) is a sufficient specification. Such a column is not strictly realizable because any flow requires in practice a pressure gradient which would cause gas properties to vary; this gradient is sometimes neglected, but in any case the growth or transition region where convection occurs is nearly always long enough to be important. However, Emmons and Land (1962) experimented with a fully-developed arc column in Poiseuille flow, for which the simple wall-stabilized model is sufficient.

In recent years several theoretical solutions have been put forward for the convection-stabilized arc. In an early analysis Suits and Poritsky (1939) regarded the arc as a vertical solid heated cylinder and used available knowledge of heat transfer in such a situation; this is not an adequate model for most purposes, but many years elapsed before a better emerged. Stine and Watson (1962) analysed a model in which the energy equation included axial convection and conduction terms, but the conducting core was supposed to maintain constant radius. Weber (1964) allowed radial convection as well as axial, and neglected axial conduction; but his solution assumes no heat conduction at the boundary of the current-carrying column, within which the Elenbaas–Heller solution is applicable [that is, the solution for no convection according to eqn (9)] and beyond which the flow is cold. More recently, Topham (1971a) has used the same equations and produced an approximate solution, computed for real gas properties, with less restrictive assumptions; and Cowley (1971) has undertaken detailed solutions of the same equations with idealized forms for gas properties.

There are several assumptions common to virtually all solutions of the arc equations for forced convection, in addition to that of thermal equilibrium. The fluid is assumed to be in axisymmetric laminar flow, and radial gradients are supposed large compared with axial (as in boundary-layer theory); the Mach number is taken to be low enough to make the kinetic energy flux negligible compared with enthalpy flux, gravity and magnetic forces are neglected, and viscous dissipation is neglected. The electric field E is taken to be purely axial, constant in r and given by J/σ, thereby causing slight infringement of current continuity and Laplace's equation. The electrical conductivity is frequently taken to be zero outside a specific isothermal, which may be in the range 4000–7000 K. Most of these assumptions have been carefully examined and many are easily defensible; others such as the basic assumption of thermal equilibrium, laminar flow and absence of magnetic forces are more restrictive.

As in other spheres of fluid mechanics, these analyses have been developed in terms of dimensionless groups, which enable universal characteristics to be derived. Reliable experimental results in forced convection are not as yet plentiful. One way to achieve quasi-steady flow is by the use of a shock-tube, of which the running time is readily made long enough to produce stationary conditions for a sufficient interval. Topham (1971a) has used such a method with nitrogen; the agreement between the measurements and his theoretical analyses can be seen from Fig. 6 which also shows Cowley's characteristic and some measurements in nozzle flow due to King (1964). The coordinates for the universal characteristics of this figure are a non-dimensional form of EI and the product of the Nusselt, Reynolds and Prandtl numbers, which together embrace I^2, gas velocity and pressure, and axial distance. The data can readily be transformed into more conventional E–I characteristics, and would then show that at large currents, other things being equal, E becomes constant in constant-pressure flow but increases with I in nozzle flow under a pressure gradient.

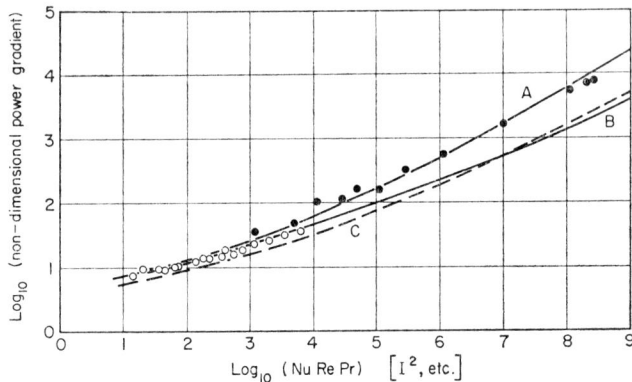

FIGURE 6. The characteristic of an arc in forced convection. A: Nitrogen flow with pressure gradient (Topham, 1971a). B: Nitrogen, constant-pressure flow (Topham, 1971a). C: Idealized gas, constant-pressure flow (Cowley, 1971). ●: Experimental points for nitrogen in nozzle flow (King, 1964). ○: Experimental points for nitrogen in constant-pressure flow (Topham, 1971a).

It is generally true that convection cooling produces very hot columns and high power gradients; within the conducting column thermal conduction predominates and the Elenbaas–Heller solution is a good approximation. The voltage gradient also rises with pressure, and values in excess of 10 kV/m are not uncommon at attainable conditions.

D. FREE-BURNING ARCS

There is a long history of investigation of the arc in free-burning atmospheric conditions, but there is to date no satisfactory model for its behaviour. It is stabilized, to a greater or lesser degree, by natural convection and therefore the momentum equation must include the buoyancy force ρg due to gravity, while the energy equation remains as for forced convection. Attempts at a solution normally postulate a conducting region and at least one outer region with artificial boundary conditions; some success has been achieved with correlating experimental characteristics (of which there are many) in terms of non-dimensional parameters. As high currents, beyond some 100 A, plasma jets may cause a substantial contribution to the convection (Section III, F).

E. THE CHANNEL MODEL

A very simple model for an axially uniform column is that of a conducting channel of radius r_c and uniform temperature T_c, losing heat by conduction across its discontinuous boundary. If both r_c and the electric field E are known, the energy equation gives T_c and the current I follows from $\sigma(T_c)$; but there is no boundary condition to fix r_c in the first instance. It is usual to make the solution determinate by invoking Steenbeck's minimum principle which asserts that E assumes a minimum value in given conditions. [This principle has been demonstrated on thermodynamic grounds by Peters (1956).] Alternatively one may require that in certain properties (such as voltage gradient, axis temperature or total radiation) the channel should match the Elenbaas–Heller solution for the same current. In most cases the channel model gives an excellent approximation to the overall behaviour of the wall-stabilized column.

F. PLASMA JETS

An important phenomenon which is usually neglected in the analyses discussed above is that of the electrode jets which emanate from both cathode and anode of high-current arcs. The reaction forces on the electrodes and the measured velocities of these jets are too large to be accounted for by vaporization rates. Maecker first explained the observed behaviour in terms of the pinch effect, which produces a pressure gradient higher at the electrode spots than in the column where the current density is much lower, and gave a convincing demonstration of this theory by producing clearly visible jets from an artificial constriction at a point in the column (Maecker, 1955).

In high-current free-burning arcs plasma jets induce sufficient convection to have a marked stabilizing effect. Topham (1971a) also points out that a plasma jet from a downstream electrode in forced convection would tend to offset the error caused by the neglect of the electrode wake in theoretical models.

G. GAS PROPERTIES

1. *Electrical conductivity*

There is no difficulty in expressing the electrical conductivity of the arc plasma formally in terms of particle densities, temperature and cross-sections, the last being also functions of temperature. If thermal equilibrium exists, the first two are related by Saha equations. The cross-sections are not known with certainty, and average values deduced from arc measurements may be used, as outlined in Section III, A. Hermann and Monterde-Garcia

(1967) discuss this in some detail. There are many published curves for the conductivity of argon and nitrogen especially, and some are based on meticulous analysis of all available theoretical and experimental information (e.g. Wells, 1967; Devoto, 1967); recently there have also been notable results for sulphur hexafluoride.

2. Thermal conductivity

The thermal conductivity of the arc fluid is determined not only by the classical flux of heat along a temperature gradient but also by the energy transport caused by the diffusion of dissociated particles (including electrons and ions) from hot to cold regions and the energy released upon their recombination. Fortunately the latter effect can be represented as a direct contribution to the total conductivity κ. Thus, ionization of an atom of mass M and ionization potential V_i comprising a fraction x of a total density ρ contributes to κ an amount

$$\kappa_i = \frac{\rho}{M} D_a(\tfrac{5}{2}kT + eV_i - \Delta\varepsilon)\frac{\partial x}{\partial T} \qquad (12)$$

at temperature T, where D_a is the ambipolar diffusion coefficient and $\Delta\varepsilon$ the amount by which the total average excitation energy of an atom exceeds that of an ion at temperature T (found from the partition functions). Similar contributions can be written for other reactions. Because in equilibrium $\partial x/\partial T$ passes a peak with increasing T, so too does the κ-contribution from any reaction. The total conductivity and its components for nitrogen as calculated by King (1956) are shown in Fig. 7.

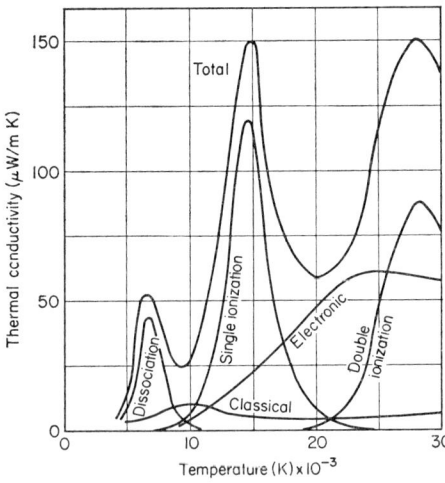

FIGURE 7. The thermal conductivity of nitrogen (after King, 1956).

ARC PHYSICS

The exception to this pattern is transport by radiation, which cannot be treated as a coefficient uniquely determined by temperature: radiation emitted in the region of a point depends only on T, but that absorbed does not. If the arc column is totally transparent to all radiation, then radiation transport is simply a separate term in the energy eqn (9) and can be accounted for independently of κ. On the other hand if the plasma is so opaque that the absorption length for all frequencies is small, then the effect becomes local and is part of $\kappa(T)$. But if, as is common, part of the radiation is absorbed in distances of the same order as the column radius, then it can be properly accounted for only by a spatial integration of absorption from all points, as well as an integration over all frequencies. Such a computation is lengthy, but for accuracy must be attempted if radiation is significant in energy transport.

The conductivity curve $\kappa(T)$ deduced from measurements in cascade arcs (Section III, A) includes this radiation transport and is therefore not the unique function sought unless a correction is made. Recent calculations have shown that previous curves should be considerably reduced in the vicinity of the ionization peaks.

The effects of peaks in $\kappa(T)$ is clearly seen in the temperature profile of the column. Figure 8 shows profiles calculated by King (1956) for nitrogen on the basis of the Elenbaas–Heller equation. The 'cores' formed can be seen, and indeed are visible in observing the column itself.

3. Radiation

As well as its role in energy transport, radiation may still contribute a significant term in the overall energy balance by virtue of the net radiation

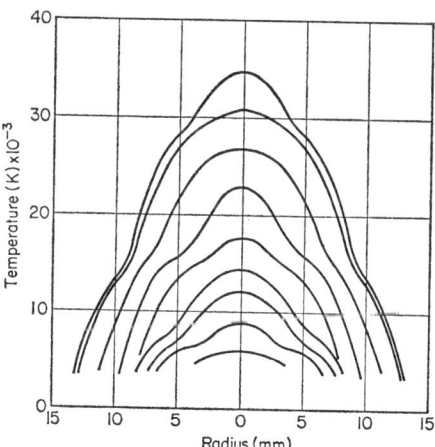

FIGURE 8. Temperature profiles for nitrogen at various currents according to the Elenbaas–Heller equation (after King, 1956).

emitted by the column. If the column is optically thin, the energy emitted per unit length is found simply by integrating $P(T)$ over the cross-section, where P is the radiated power per unit volume at all frequencies; otherwise the net emission must be found by the integrations described in the previous section. Unless pressure and current are very high, radiation is not a large part of the energy balance; according to Wells (1967) radiation from a nitrogen arc at atmospheric pressure with a power gradient of 0.9 MW/m accounts for 4% of the energy input. But Hermann and Schade (1971) find a 40% loss for nitrogen at 12 atm and some 25 MW/m.

4. *Turbulence*

There is no doubt that in many arcs with convective heat transfer there is some degree of turbulent flow. In the conducting column the flow is thought normally to be laminar, but there may be turbulence in other regions which would affect heat transport properties. There is at present an upsurge of interest in this problem; but turbulent heat transfer is a complex phenomenon even in otherwise simple situations, and its introduction into the physics of the arc column has so far been semi-empirical. In any case, theories which assume laminar flow have not been discredited by experiment; and in those cases where there is a pronounced discrepancy it is not safe to attribute this to turbulence before other sources have been eliminated.

H. THE DYNAMIC BEHAVIOUR OF THE COLUMN

The general forms of the equations for continuity, momentum and energy are time-dependent and must so be used in any rigorous analysis of the time-varying arc column. If convection and radiation are ignored, the energy equation becomes

$$\sigma E^2 + \nabla . \kappa \nabla T = \rho \frac{\partial h}{\partial t}, \qquad (13)$$

in which E is a function of time t. The solution is complicated by interaction between arc and circuit, and may be somewhat simplified by taking current to be, as before, an independent variable (that is determined only by the circuit). Phillips (1967) solved eqn (13) for sinusoidal current in a column with constant diffusivity $\kappa/c_p\rho$ and a linear $\sigma(T)$ function. He considered two cases: the low-frequency case where the radius varied with time and the E–I relationship was virtually that for the static arc; and the high-frequency case where the radius is virtually constant and the E–I characteristic approaches linearity.

There is a long history of investigation on the decay of the column after its current falls to zero (Edels and Ettinger, 1962; Jones and Edels, 1969;

Hertz and Maecker, 1971). The agreement between theory and experiment is in some cases unsatisfactory for molecular gases; their rates of recombination may be too slow to maintain approximate LTE.

Two early and quite simple models were due to Cassie (1939) and Mayr (1943). The first was based on a channel model in which only the radius varies with time; Mayr assumed a constant radius and a varying temperature with an energy balance by conduction only in an idealized gas. Browne (1948) demonstrated that for a continuous current change Cassie's model was a good approximation at high current, while Mayr's was appropriate to the region where the current approached zero. Both models imply that the column decay is governed by a single time-constant, which is not the case either in experiment or in more rigorous analysis. Frind (1960) solved eqn (13) for a constant diffusivity and a linear law between electrical conductivity and heat-flux potential; he deduced a time-constant

$$\tau = \tfrac{1}{2}\left(\frac{r_c}{2.4}\right)^2\left\{1 - \frac{d(\ln E)}{d(\ln I)}\right\}, \tag{14}$$

in which r_c is the conducting radius. With a falling E–I characteristic (as at relatively low currents) the last factor is not very different from unity, and τ then varies as the cross-sectional area of the conducting column.

Topham (1971b) has used a dynamic model for an arc in forced convection, with a pressure gradient chosen to give axial uniformity in the current-carrying column and with viscosity neglected. An approximate solution of the energy equation, computed for real gas properties, gives a non-dimensional voltage–time curve for each chosen current variation, in variables which include the effect of gas velocity and pressure; not only does this model combine the features of the Cassie and Mayr equations, but unlike them it proceeds to a correct steady-state limit.

IV. The Electrode Regions

In referring to the electrode regions, we usually mean the very short regions of high space-charge across which the anode and cathode falls of potential are developed. The transition between these and the column is not sharp, and there are contraction regions which have column-like properties and in which the current density changes from its very high value at the electrode surface to the distribution determined by the boundary conditions of the conducting column. There have been theories which embrace the contraction regions as well as the electrode region proper, notably by Ecker (1953), but for the most part only the latter have been considered to be radically distinct from the column. The cathode has received more attention than the anode, because its emission is essential to the passage of current.

Guile (1971) has recently reviewed published work on the electrode phenomena of arcs. There is a very extensive corpus of both experimental data and theories; unhappily neither the one nor the other displays as yet any satisfying coherence and the physics of electrode processes remains obscure. The obscurity is almost literal: the experimenter is faced with regions of microscopic dimensions at temperatures and levels of brightness which almost rule out instrumental measurements and in which there occur at once electrical, thermal and mechanical processes. It is to be wondered at that we know as much about it as we do.

A. THE CATHODE REGION

There are two main categories of arc cathode behaviour: the thermionic cathode in which the temperature can be high enough to produce the arc current by thermionic emission, and the non-thermionic cathode in which one or more other mechanisms must act. Thermionic cathodes (which may, of course, be externally heated but which here we suppose to be heated by the arc itself) are those made of refractory materials like carbon or tungsten, whose boiling points are high enough to provide copious emission. Such cathodes show current densities in the range 10–100 A/mm^2; the spot is quite stable, but may move slowly on the cathode surface.

The non-thermic cathodes are those of materials with boiling points too low to produce sufficient thermionic emission; here current densities are in excess of 10^4 A/mm^2 and the spots are unstable, moving rapidly and erratically. The emission is on the whole believed to be by a form of field emission, and to explain such current densities one needs fields in the order of 10^6 V/mm.

The length d_c of the space-charge region, its voltage V_c (the cathode fall) and the current density J_c may be related by the Child–Langmuir equation which gives

$$J_c = \frac{4\sqrt{2}}{9} \varepsilon_0 \sqrt{\left(\frac{e}{m}\right)} \frac{V_c^{\frac{3}{2}}}{d_c^2}, \tag{15}$$

where e/m is in this case the charge–mass ratio of the positive ions. V_c can be measured with relative ease, although the contribution from the contraction region is difficult to exclude; but J_c cannot be measured with any accuracy and eqn (15) makes the assumptions that ions enter the region with zero velocity, and that electrons in passage do not influence the field. So it is not possible to deduce a reliable value for d_c; nevertheless, eqn (15) and such experimental evidence as exists (Dickson and von Engel, 1967) indicate that d_c is in the range 10^{-5}–10^{-1} mm. It is often suggested that d_c is in the order of a mean free path, which is not inconsistent with these figures, but the value of any mean free path is itself uncertain in the region of the electrodes. A consequence of eqn (15) is that the electric field at the cathode is

$$E_c = \frac{4V_c}{3d_c},$$

which gives values in the order of 100–10^6 V/mm; just enough, that is, to account for the non-thermionic cathode on the basis of ordinary field emission. But the margin is too fine for comfort, and calculations for particular arcs frequently indicated that field emission is not sufficient (especially at a time when current densities were thought to be lower than now). It has therefore been common to suggest that the field emission is enhanced by surface defects and by layers of oxide or contaminants on which a layer of positive ions can form. Certainly oxidation and contamination are known to affect cathode behaviour; for example, they encourage the glow-to-arc transition (Section I, C) and make the arc less liable to extinction. But there is little direct evidence of their role.

Thus, whereas the thermionic cathode process appears to be well-confirmed, there is no such certainty about the cathode of the non-thermionic arc (also called the vapour arc or, misleadingly, the cold-cathode arc). Other theories have been put forward for the latter which include almost every conceivable way of providing the energy to overcome the cathode work function. Among them is the suggestion by von Engel and Robson (1957) that excited atoms are scattered on to the cathode surface and give up their energy at a sufficient rate to account for the emission; it is supposed that vapour pressure at the cathode spot is much higher than ambient, perhaps by a factor of 10.

Although thermionic emission alone has been ruled out for the 'cold' cathode it is established that refractory cathodes can support non-thermionic cathode spots. An erratic spot of very high density can be obtained, especially if the surface is contaminated, and the transition to this mode from the thermionic can be shown by lowering the gas pressure: the current density of the spot, which in the thermionic process decreases with falling pressure, rises again sharply at pressures in the region of 0.1 bar.

The cathode fall of potential is of the same order as the ionization potential of the arc gas; it accelerates electrons sufficiently to produce intense excitation and ionization (in which cumulative ionization may be important) where the space-charge region ends. Ions so produced are accelerated towards the cathode surface and there give up energy as heat. The contribution of ions to the current may therefore be greater at the cathode than elsewhere, but it is probably still very small.

It is possible to draw up an energy balance for the cathode region. For every unit of current flowing, of which a fraction x is contributed by positive ions at the cathode surface, the cathode receives electrical power

$$W_e = x(V_i + V_c) - \phi_c \qquad (16)$$

in which V_i is the ionization potential of the gas and ϕ_c the cathode work function. Heat is lost from the cathode surface by conduction, radiation and evaporation or ablation. In each of these the net loss is the difference between heat lost and heat gained, since the cathode receives heat by conduction and radiation from the column, and by the condensation of some ions which are not reflected as neutrals. The overall heat loss is then balanced by W_e. The conduction loss from the cathode spot can be estimated by regarding it as a circular source of known radius and temperature in a semi-infinite medium, and net evaporation rates can be estimated from experiment; but there remains uncertainty as to the value of x, and the energy balance is therefore used more to eliminate the impossible than to provide positive answers.

The problem of the non-thermionic cathode spot is made more difficult not only by rapid random movement, but also by the fact that above some value of current the spot splits into separate emitting sites (which are very close together at high pressures but at low pressures may be widely spaced, as in the vacuum arc). There is no precise critical value for splitting, and the current in each site can vary from a fraction of an ampere upwards depending on the conditions. The sites do not exist continuously but have an average lifetime, of which estimates are as short as 10^{-9}s (Guile, 1971), with a balance between generation and decay. There is evidence that the emission from a given site collapses when oxide or impurity layers have been removed by bombardment; and that on a polycrystalline surface sites tend to concentrate on grain boundaries or on particular crystal planes.

At atmospheric pressure, arcs are usually influenced by external magnetic fields in predictable ways but at lower pressures the phenomenon of retrograde motion may be observed, in which the cathode spot moves in the direction apparently opposite to that dictated by electrodynamics; so far as is known, there is no retrograde force on the column. Retrograde motion is favoured not only by low pressures but also by high magnetic fields and short columns (it has been observed at atmospheric pressure for arcs less than 2 mm long carrying currents below 5 A in a field exceeding 0.3 T: von Engel and Robson, 1956). At pressures below about 1 mbar the current limit is quite high, e.g. some hundreds of amperes. There is as yet no satisfactory explanation of the effect, although it boasts an extensive literature.

B. THE ANODE REGION

The anode region has much in common with the cathode: it merges with the column by way of a contraction zone and has a very short region of space charge, negative in this case, responsible for a potential drop in the same order as (but usually rather less than) the cathode fall. The current density at the anode surface, similarly, is normally higher than in the column but less than at the cathode: it lies in the range 1–1000 A/mm^2.

Unlike the cathode fall, the anode fall cannot be explained as an essential part of an emission process. There are two common views of its function. One holds to the Langmuir-probe analogy: if the anode must receive an electron current density exceeding the random value in the column, then the anode must assume a potential positive to the column plasma and an electron sheath forms accordingly. This view is certainly plausible at lower pressures, to which conventional probe theory is readily applicable. The other view takes the anode fall to be a means of producing sufficient positive ions to supply the end of the column. Since, nearly always, the anode emits no ions the density of positive ions must be built up in the anode region. A strong electric field can do this in a short space, either by acceleration through a short collisionless layer or by enhancing electron temperature; both mechanisms have been proposed.

It remains to explain why the anode region requires higher current densities than the column despite the circumstances that no emission is required and the anode area is rarely a limitation. Usually Steenbeck's minimum voltage is invoked: the total potential drop is lowered either by the current concentrating on a point of low work function, or by it concentrating sufficiently to produce enough metal vapour to lower the local electrical conductivity. Certainly anode spots are very hot and there is intense vaporization; there is also evidence of several constituent spots although in general the spot shows no random motion comparable to that shown by non-thermionic cathodes.

The energy balance of the anode surface follows that at the cathode, but is simpler to the extent that the electrical input per unit current is simply $V_a + \phi_a$. The velocity of a vapour jet emanating from the anode can be at least as great as that from the cathode, even although the current density and therefore the pinch effect are less; so it appears that a contributing factor to these jets is intense vaporization and high vapour pressure.

V. Arc Measurements

A. ELECTRODE MEASUREMENTS

The measurements most frequently attempted for cathode or anode spots are those of temperature and size. The only direct way of measuring temperature is by pyrometry, which is more readily applied to stationary spots (where there is considerable luminescence) than to moving. There have been very many measurements of spot size, aimed at determining current densities: for this the two available methods are direct observation and the examination of marks left on surfaces after arcing has occurred. Both are unreliable. The apparent visual size, however the emitted light is filtered, may not be that at the actual surface, nor that which carries current, nor that of only one spot. High-speed photography helps to resolve the last objection and is also used to measure the multiplicity of spots and their velocity, in the non-thermionic cathode. Measurement of marks is also confused by the effect of several close spots, and there is no certainty of how the eroded or ablated area is related to the current-carrying area.

A measurement which can be done with fair accuracy is that of the electrode falls of potential. If the electrode spacing is decreased steadily, the arc voltage falls steadily (reflecting the change in column voltage) until the spacing is very small, when the voltage falls suddenly to nearly zero. This drop in voltage can be safely taken to be the sum of the cathode and anode falls; with fast recording it is possible to separate the two (Dickson and von Engel, 1967) for the anode fall disappears shortly before the cathode fall, which is the last essential in supporting the arc before it is short-circuited by the touching electrodes. A less accurate way to estimate the falls is to extrapolate the linear (or nearly linear) portion of a curve of total voltage against gap length, obtained by measuring voltages for arcs of different lengths in identical conditions; this fails to account for the inevitable contraction region adjacent to each electrode, and consequently gives values which are too high.

Dickson and von Engel also used the moving electrode experiment to estimate the lengths of the fall or space-charge regions from a knowledge of velocity and the duration of the voltage steps.

B. COLUMN MEASUREMENTS

A large proportion of experimental work with arcs aims to measure properties of the column, supporting the theoretical models which are available for different situations and which are quite highly developed. Among the more refined measurements on the column are those of temperature and density; these we deal with separately.

1. *Temperature and density*

There is no direct, that is thermometric, way of measuring the high temperatures of the column. Available methods are of two basic kinds: those which measure gas density from which temperature follows by an equation of state, and spectroscopic methods which usually measure populations or distributions of radiating species from which temperature follows from the statistics of equilibrium.

There are several methods of finding the gas density. Early measurements of column temperature were made by von Engel and Steenbeck (1931, 1933) using the change in the absorption coefficient of soft X-rays or of α-particles to deduce the density on the axis. Soon after, Suits (1935) measured the velocity of sound along the axis, a method which Edels and Whittaker (1957) applied more rigorously to the velocity of the shock wave generated by a spark. The density may also be measured by way of the refractive index, a method applicable mainly to the cooler outer regions of the column. The change in refractive index with density is recorded as a fringe pattern by interferometric, Schlieren or shadowgraph techniques. The difficulties of using a light source in the presence of the highly luminous column can be overcome by using laser light and narrow-band filters. Such methods are helpful in giving a qualitative view of the thermal field.

Of the many ways of measuring temperature spectroscopically, one of the most often used is that of excitation temperature by relative line intensity. If the transition probability of a line is known, together with its energy levels and their statistical weights, then the ratio of the intensity of two lines gives the ratio of populations for their upper states, and from this the temperature follows by the Boltzmann law, eqn (4). This assumes not only that there is a near enough approach to equilibrium for the law to be applicable, but also that the plasma is collision-dominated and optically thin; the method is therefore (like most spectroscopic methods) subject to the reservations discussed in Section II. It is also of doubtful accuracy if the energy difference in the Boltzmann factor is small, and is best applied to a series of lines; a logarithmic plot of I/Agf against energy, where I, A, g, f are the intensity, transition probability, statistical weight and frequency, should give a line of slope $1/kT$. A similar method uses the intensity distribution within a rotational band of a molecular spectrum; here it is usual to plot the intensity variable in terms of the rotational quantum numbers of the lines, which are a measure of their statistical weights.

Temperature may also be deduced from the absolute intensity of a line, given in equilibrium by

$$I = \frac{Agfnh}{4\pi b} e^{-\varepsilon/kT} \qquad (17)$$

per unit volume and steradian, where b is a partition function, ε the energy and n the number density of the radiating species. This method can be accurate provided that n and b are reliably known and that I can be accurately measured by means of a standard source. The density n itself depends on temperature, and may be inserted as a function of T or separately measured.

Another method which uses a chosen line needs no calibration. Since the population of any particular species shows a maximum at a certain temperature (e.g. Fig. 4) the intensity of any line also shows such a maximum, at a temperature which is calculable. If the hottest parts of the column exceed such a value, then a point of known temperature can be located by observing the line intensity; the temperature of other points can then be deduced from the variation of intensity.

Other properties of individual lines can be used. If self-absorption of a line is very strong, its intensity may reach the black-body limit, evidenced by a truncated profile; the temperature then follows from Planck's law. If the spectrum is observed in the direction of a temperature gradient (e.g. transverse to the axis) self-reversal can be seen if the absorption is strong, and it is possible to deduce temperatures from this. Doppler broadening of a line profile also gives gas temperature: in principle this is a simple method which depends only on a Maxwell distribution for the species concerned and not on Boltzmann populations; but unfortunately the Doppler broadening is frequently obscured by the other causes of line broadening. Wulff (1958) made a detailed study of the relative importance of different broadening processes in a helium arc.

The continuum produced by radiative electron–ion recombination and free–free transitions (bremsstrahlung) affords estimates of electron density n_e and temperature T_e if the absolute intensity distribution is measured. In general, the dependence of the continuum intensity on T_e and n_e is complicated (even more so if radiative attachment is also important) but the temperature can be deduced from the distribution in wavelength; the absolute intensity is in many cases closely proportional to n_e^2.

A more straightforward basis for measuring n_e is the Stark effect, that is the broadening of spectral lines by the action of the electrical fields of surrounding charge carriers on the emitting particles. It can be applied only if other broadening sources are satisfactorily accounted for. In hydrogen, Stark broadening is particularly pronounced and a hydrogen line profile gives a direct measure of $n_e^{2/3}$.

The recording of spectra for these measurements may be done either by photomultipliers or by photographic plates; the former is accurate and makes time resolution much easier, while the latter is better suited to spatial (and spectral) resolution in stationary columns (if a certain region of the column is focussed on the entrance slit, the recorded spectrum is the image of this region). In both cases, however, the spectrum is made up of radiation emitted from all points lying along a line of sight. If the column is viewed from a direction transverse to its axis, as when examining radial temperature profiles, the radiation recorded is integrated along a chord of the cross-section. A standard problem in deducing profiles, therefore, is to convert the variation of recorded intensity from chord to chord into the desired variation with radius. The conversion requires the solution of an Abel integral equation; there are various procedures readily available for doing this, but they must be done with care to avoid gross errors (Lochte-Holtgreven, 1968).

2. *Other measurements*

The axial voltage gradient in the column may be estimated from a curve of arc voltage against length; this assumes not only that the gradient is uniform (experimental curves usually indicate uniformity over a range of length, even where the column is not uniformly bounded), but also that the column properties are independent of total length. To avoid large changes in length, some experimenters have used an electrode vibrating with known amplitude; the amplitude of the voltage modulation then gives the field, subject to the above assumptions.

Although the theory and practice of electrostatic probes at high pressures are still ill-matched, the voltage difference between two points in identical plasma conditions (e.g. axially separated points at the same radius) can be found from probes at these points. Provided only that the plasma-to-probe potential is the same at each, the probes record the potential difference between the points in the plasma. This provision requires first that the plasma is identical in nature at each probe, and secondly that the probe potentials are adjusted until each draws zero current (or a small current which is the same for both). In the case of the cascade arc (Section III, A), which has been used for many of the most careful arc measurements, two of the cascade elements can be used as probes, with the advantage that they present no disturbance to the symmetry of the column.

It has been usual to estimate the radius of the electrically conducting column photographically, by taking it to be the luminous radius, although it is clear that the outer non-conducting regions must contribute to visible radiation. Siddons (1971) has compared the luminous diameters of arc columns with the conducting diameter measured by a fast-moving probe, biased to draw current, whose position was tracked photographically. He found that in a nitrogen arc carrying 1200 A in forced convection the conducting diameter was on average about 0.8 of the luminous diameter.

In the case of arcs with convection, it is desirable to know the gas velocity in the region of the column. Local velocity may be measured by a tracer technique: Wienecke (1955) used both small sparks (in the central regions) and carbon particles (in the outer regions) to map the complete velocity fields of a high-current carbon arc in which convection, with velocities over

200 m/s, was induced by the cathode jet (Section III, F). No such measurements have been reported as yet for the arc in forced convection, but here the velocity of the cold gas is the important parameter, and this is usually known independently.

References

Ayrton, H. (1902). "The Electric Arc". The Electrician Publishing Co., London.
Browne, T. E. (1948). *Trans. Am. Inst. elect. Engrs.* **67**, 141.
Burhorn, F. (1959). *Z. Phys.* **155**, 42.
Burhorn, F., Maecker, H. and Peters, Th. (1951). *Z. Phys.* **131**, 28.
Cassie, A. M. (1939). *Conf. int. gr. Réseaux élect. hte. Tens.* 10, Paris.
Cowley, M. D. (1971). Cambridge University Engineering Dept. Report A-ARC/TR4.
Devoto, R. S. (1967). *Physics Fluids* **10**, 354.
Dickson, D. J. and von Engel, A. (1967). *Proc. R. Soc.* A **300**, 316-325.
Drawin, H. W. (1970). *Z. Naturf.* **25a**, 145-147.
Druyvesteyn, M. J. and Penning, F. M. (1940). *Rev. mod. Phys.* **12**, 87.
Ecker, G. (1953). *Z. Phys.* **136**, 1.
Edels, H. (1961). *Proc. Instn. elec. Engrs.* **108A**, 55-69.
Edels, H. and Ettinger, Y. (1962). *Proc. Instn. elec. Engrs.* **109A**, 89-98.
Edels, H. and Whittaker, D. (1957). *Proc. R. Soc.* A **240**, 54-66.
Emmons, H. W. and Land, R. L. (1962). *Physics Fluids* **5**, 1489-1500.
Frind, G. (1960). *Z. angew. Phys.* **12**, 231-237.
Gerdien, H. and Lotz, A. (1923). *Z. tech. Phys.* **4**, 157-162.
Guile, A. E. (1971). *Proc. Instn. elec. Engrs.* **118**, 1131-1154.
Hermann, W. and Monterde-Garcia, A. (1967). *Z. Phys.* **205**, 313-327.
Hermann, W. and Schade, E. (1971). *Int. Conf. Ioniz. Phenom. Gases* 10, Oxford.
Hertz, W. and Maecker, H. (1971). *Int. Conf. Ioniz. Phenom. Gases* 10, Oxford.
Jones, G. R. and Edels, H. (1969). *Z. Phys.* **229**, 14.
King, L. A. (1956). *Int. Conf. Spectrosc.* 6, Amsterdam.
King, L. A. (1964). *Rep. Br. elect. all. Ind. Res. Ass.* 5072.
Lochte-Holtgreven, W. (1968). *In* "Plasma Diagnostics" (W. Lochte-Holtgreven, ed.) pp. 135-213. North Holland, Amsterdam.
Lord, W. T. (1967). *Rep. R. Aircr. Establ.* TR 67087.
Maecker, H. (1955). *Z. Phys.* **141**, 198.
Maecker, H. (1956). *Z. Naturf.* **11a**, 457-459.
Maecker, H. (1964). *In* "Discharge and Plasma Physics" (S. C. Haydon, ed.) pp. 245-265. University of New England, Australia.
Mayr, O. (1943). *Arch. Elektrotech.* **23**, 588-608.
McWhirter, R. W. P. (1963). *In* "Plasma Diagnostic Techniques" (R. H. Huddlestone and S. L. Leonard, eds.) pp. 201-264. Academic Press, New York and London.
McWhirter, R. W. P. and Hearn, A. G. (1963). *Proc. Phys. Soc.* **82**, 641-654.
Peters, Th. (1956). *Z. Phys.* **144**, 612.
Phillips, R. L. (1967). *Br. J. appl. Phys.* **18**, 65-78.
Richter (1971). *Int. Conf. Ioniz. Phenom. Gases* 10, Oxford.
Rompe, R., Thouret, W. and Weizel, W. (1944). *Z. Phys.* **122**, 1.
Siddons, D. J. (1971). Thesis, Oxford University.

Stine, H. A. and Watson, V. R. (1962). *NASA Rep.* TND 1331.
Suits, C. G. (1935). *Physics* **6**, 315.
Suits, C. G. and Poritsky, H. (1939). *Phys. Rev.* **55**, 1184-1191.
Topham, D. R. (1971a). *J. Phys.* D **4**, 1114-1125.
Topham, D. R. (1971b). *Int. Conf. Ioniz. Phenom. Gases* 10, Oxford.
Uhlenbusch, J., Fischer, E. and Hackmann, J. (1970). *Z. Phys.* **239**, 120.
von Engel, A. and Robson, A. E. (1956). *Phys. Rev.* **104**, 15-16.
von Engel, A. and Robson, A. E. (1957). *Proc. R. Soc.* A **243**, 217–236.
von Engel, A. and Steenbeck, M. (1931). *Wiss. Veröff. Siemens-Werken* **10**, 155.
von Engel, A. and Steenbeck, M. (1933). *Wiss. Veröff. Siemens-Werken* **12**, 74.
Wasserab, T. (1950). *Z. Phys.* **127**, 324.
Weber, H. E. (1964). *AGARDograph* **84**, 845-881.
Wells, A. (1967). *Rep. R. Aircr. Establ.* TR 67076.
Wienecke, R. (1955). *Z. Phys.* **143**, 128.
Wulff, H. (1958). *Z. Phys.* **150**, 614.

Gravitation

WALTER THIRRING

Institute for Theoretical Physics, University of Vienna, Vienna, Austria.

I. Introduction	125
II. Universality	127
III. Weakness	130
IV. Thermodynamic behaviour of gravitating particles	134
V. Life and death of stars	141
VI. Gravitational radiation	146
VII. The equivalence principle	147
VIII. Strong and weak equivalence	152
IX. A trip behind the Schwarzschild radius	156
References	161
Appendix	161

I. Introduction

The problem of gravitation was solved by Einstein in 1916. However, in the meantime our knowledge of particles and their interactions has been increased to such an extent that a discussion of the subject looked at from the point of view of particle physics seems worthwhile. In particular, in this essay I shall give an elementary discussion of theories of gravitation constructed according to the pattern of classical electrodynamics. (Rohrlich, 1965) In this the equations of motion of particles and the field equations are derived from one action principle. The total action W is the sum of a contribution from the particles W_{part}, one from the field W_{field}, and they are coupled by an interaction term W_{int}

$$W = W_{part} + W_{field} + W_{int}. \qquad (1.1)$$

In the covariant description where the four co-ordinates z^i, (z^0 = time, $z^{1,2,3}$ = space), are considered as functions of the proper time s, W_{part} can be written

$$W_{\text{part}} = \frac{m}{2} \int ds\, \dot{z}^i(s)\, \dot{z}^k(s)\, \eta_{ik}. \tag{1.2}$$

Here $\dot{z} = dz/ds$ and η is the metric tensor.

$$\eta = \begin{pmatrix} 1 & & & \\ & -1 & & \\ & & -1 & \\ & & & -1 \end{pmatrix}$$

and we have written the contribution of one particle to W.

For several particles it is the sum thereof. With the vector potential A_i and its derivatives $A_{i,k}$ the remaining parts of W are

$$W_{\text{field}} = \frac{1}{4} \int d^4 x (A_{i,k}(x) - A_{k,i}(x))(A_{e,m}(x) - A_{m,e}(x))\, \eta^{ie} \eta^{km}, \tag{1.3}$$

$$W_{\text{int}} = e \int ds\, A_i(z(s))\, \dot{z}^i(s). \tag{1.4}$$

Thus, the theory depends on two parameters, the inertial mass m and the electric charge e.

There are some obvious possibilities of constructing similar covariant theories. For instance, with a scalar potential ϕ or a tensor ψ_{ik}, we can construct the invariant interactions

$$W_{\text{int}}^{(s)} = g_s \int ds\, \phi(z(s)), \tag{1.5}$$

$$W_{\text{int}}^{(T)} = g_T \int ds\, \psi_{ik}(z(s))\, \dot{z}^i(s)\, \dot{z}^k(s). \tag{1.6}$$

For the scalar theory we could use

$$W_{\text{field}} = \tfrac{1}{2} \int d^4 x\, \phi_{,i}(x)\, \phi_{,k}(x)\, \eta^{ik} \tag{1.7}$$

whereas for the tensor case there are five possible terms quadratic in the derivatives of ψ. All these theories give in the non-relativistic limit a potential between two particles $\sim 1/r$ and differ only in their relativistic features. The theories in the framework indicated above will be sufficient for our discussion of gravitation.

The most important feature of gravitation is its universality which goes much beyond that of other interactions. It appears in the legendary vision of Newton under the apple tree where he sensed that everything, we, the apple, the moon, are subject to the same universal force. In Section II we shall review how our experimental knowledge on this point has evolved. In fact, elementary particle physics has provided a fantastically sensitive instrument which allows us to verify this fundamental property of gravity with a precision of at least 1 in 10^{14}. In Section III we come to the other distinguishing feature of gravitation, its extreme weakness. In fact, whereas $e^2/\hbar c \sim 10^{-2}$, we have for gravitation $g^2/\hbar c \sim 10^{-38}$. Nevertheless, due to its universality gravitation dominates the scene if sufficiently many particles are around. First of all, it introduces unusual features in thermodynamics. One finds that non-relativistic fermions (i.e. spin $\frac{1}{2}$ particles) with attractive $1/r$ interactions have a region of negative heat capacity in the microcanonical ensemble which in the canonical ensemble is bridged by a phase transition. This tendency of gravitating matter to give away energy and thereby to contract and to become hotter is discussed in Section IV. In real stellar evolution the situation may become more dramatic once the particles become relativistic. Then the Fermi pressure is no longer sufficient to stop the contraction and the phase transition finds no end. In Section V we enter the relativistic region and then other effects which did not appear in the static picture are no longer negligible. In Section VI gravitational radiation is considered and in Section VII we come finally to that consequence of gravitation which was the most sensational aspect of Einstein's theory: it deforms the geometry of space–time! We shall see that this is indeed a consequence of theories with universal interactions of the type (1.5) or (1.6). Whereas under normal circumstances these effects are ridiculously small, in the extreme situations of gravitational collapse they become dominating. They lead to unusual and at first sight even paradoxical consequences which will be studied in Section VIII. Clearly, we shall not attempt to cover in this short essay all aspects of gravitation but we shall limit ourselves to these important features which are not too far away from our experimental knowledge.

Although we do not follow in our simple exposition the mathematically advanced road of Einstein, we will see that any reasonable approach will end up at Einstein's general theory of relativity. It was the culmination point of physics up to then and at the same time not only undermines some of the foundations of classical physics, but finally leads to the destruction of matter in a black hole.

II. Universality

We shall define universality in the sense of Galileo's experiment by requiring that the trajectory of a particle in an external gravitational field depends only on the initial conditions of its motion. In particular, it should be independent of its mass which means that the coupling constant g of a particle is proportional to its inertial mass. If we have several particles labelled by an index α, then there should be a universal constant f such that

$$g_\alpha = fm_\alpha. \tag{2.1}$$

Indeed, in this case m_α is a common factor of $W_{\text{part}} + W_i$ and therefore cancels in the equation of motion of the particle in the external field. Thus, the trajectories will be the same for identical initial conditions.

The first precision measurement of the equality of the inertial mass appearing in (1.2) and the gravitational mass appearing in (2.1) was made by Eötvös in his famous experiment (Eötvös et al., 1922). There the resultant of the centrifugal force due to the rotation of the earth and the gravitational force of the earth is measured. Since the former is proportional to the inertial mass and the latter proportional to the gravitational mass the direction of the resultant force should be different for particles having different ratios between the two masses. This experiment was later refined by Dicke (1964) who demonstrated the universality of this ratio with an accuracy of one part in 10^{11}. This already excluded some speculations that the gravitational mass of antiparticles would have the opposite sign of the one for particles, in analogy to the electric charge. This proposal was actually already ruled out by the deflection of light in the gravitational field since the photon, being its own anti-particle, should have no gravitational interaction according to this hypothesis. Dicke's experiment gave a more precise number since the virtual electron and positron pairs contribute a fraction of 10^{-7} to the mass of matter (Schiff, 1959). The most precise figure on the equality of the gravitational mass for particle and anti-particle comes from the system of neutral K-mesons following a proposal by Good (1961). There, one has a neutral meson K^0 and its anti-particle $\overline{K^0}$. They decay, however, in the quantum mechanical components

$$\begin{rcases} |K_S\rangle \simeq [|K^0\rangle + |\overline{K^0}\rangle]/\sqrt{2} \\ |K_L\rangle \simeq [|K^0\rangle - |\overline{K^0}\rangle]/\sqrt{2} \end{rcases} \tag{2.2}$$

up to some arbitrary phase convention, and where we have neglected the small (10^{-3}) CP-violation contribution (Marshak et al., 1969). K_S decays after about 10^{-10} sec mainly into two pions, whereas K_L decays after about 10^{-8} sec into three pions and other things. Also their inertial mass is slightly different and the difference Δ is of the order of the difference of the imaginary parts of the masses corresponding to the decay times. Therefore, Δ/m is of the order 10^{-14} which gives the system its fantastic sensitivity. A small difference in the gravitational masses of K^0 and $\overline{K^0}$ could introduce transitions between K_L and K_S and thereby completely change the decay pattern of the system. To put these ideas into equations, we consider inertial and gravitational mass as 2×2 matrices operating in the Hilbert space of the quantum mechanical $(K^0, \overline{K^0})$-system. If the gravitational mass of K^0 and $\overline{K^0}$ were slightly different, this matrix would be in the $(K^0, \overline{K^0})$ basis of the form

$$m\begin{pmatrix} 1+\delta & 0 \\ 0 & 1-\delta \end{pmatrix}.$$

The inertial mass on the other hand is given, on the same basis, by

$$\begin{pmatrix} m & \Delta/2 \\ \Delta/2 & m \end{pmatrix}$$

We now insert these expressions into the action governing the behaviour of the neutral K-system in an external gravitational field, i.e. $W_{\text{part}} + W_{\text{int}}$. This means that the translational degrees of freedom of the system are treated classically and only the internal particle–antiparticle degree of freedom is treated quantum mechanically. For W_{int} we choose the form (1.6) with g_T proportional to the gravitational mass. For the gravitational potential ψ_{ik} we use the form

$$g_T\psi_{ik} = \begin{pmatrix} -V & & & 0 \\ & -V & & \\ & & -V & \\ 0 & & & -V \end{pmatrix}, \qquad (2.3)$$

where V is the gravitational potential $V = Mf^2/4\pi R \ll 1$ in units of $c = 1$. This corresponds, according to the evidence to be discussed later, to the ψ_{ik} experienced on the surface of the earth. The total Lagrangian is then given by the matrix

$$L = \tfrac{1}{2}\begin{pmatrix} m & \Delta/2 \\ \Delta/2 & m \end{pmatrix}(\dot{z}_0^2 - \dot{z}^2) - Vm\begin{pmatrix} 1+\delta & 0 \\ 0 & 1-\delta \end{pmatrix}(\dot{z}_0^2 + \dot{z}^2). \qquad (2.4)$$

Diagonalizing Eqn (2.4) for $V\delta \ll \Delta/m \ll 1$, we find for the eigenvectors†

$$\begin{aligned} |K_S\rangle &= [|K^0\rangle + |\overline{K^0}\rangle + \varepsilon(|K^0\rangle - |\overline{K^0}\rangle)]/\sqrt{2} \\ |K_L\rangle &= [|K^0\rangle - |\overline{K^0}\rangle - \varepsilon(|K^0\rangle + |\overline{K^0}\rangle)]/\sqrt{2} \end{aligned} \qquad (2.5)$$

$$\varepsilon = \frac{m\,\delta V}{\Delta}[\dot{z}_0^2 + \dot{z}^2], \qquad (2.6)$$

and we have used $\dot{z}_0^2 - \dot{z}^2 = 1$. We note that when V or δ go to zero, we obtain the familiar eigenstates of CP, the K_1^0 and K_2^0. The parameter ε measures the famous admixture of 2π decay in the K_L state, which is known

† ε is somewhat different from the usual ε-parameter.

experimentally to be

$$|\varepsilon| \sim 2 \times 10^{-3}.$$

The important part is to recognize that when Eqn (2.6) is written in terms of the ordinary velocity v of the K-meson (say in the laboratory),

$$\varepsilon = (2m\delta V/\Delta)(1 + v^2)/(1 - v^2) \simeq 4\gamma^2 m\delta V/\Delta,$$

that is, the 2π decay ratio of the K_L should exhibit a γ^2 dependence as the K_L energy is increased. The γ^2 dependence is typical of our tensor field ($J = 2$) and is in general of the type γ^J. Thus we have no energy dependence in a scalar theory, and a simple γ dependence for the vector field.

There are many measurements (Particle Data Group, 1971) in the range 1 to 10 GeV and no γ dependence is observed, so that CP violation with a tensor field is excluded. The possibility of an interaction with a scalar field (not necessarily gravitational) is, however, still open as an explanation of the CP violation.

From the absence of the γ^2 dependence, we conclude that the 2π decay of K_L is due to an interaction other than the gravitational one. Thus we can obtain a very accurate upper limit on δ. Experimentally, from measurements up to 10 GeV/c, we obtain $|\varepsilon| = 2 \times 10^{-3}$ with an error of 10%. The value of mV that we will use depends on how large we assume the range of the gravitational interaction to be; for example, if we only include the contribution of the sun to V, $mV = 6$ eV, for the entire galaxy $mV = 500$ eV, and for the whole universe $mV = 500$ MeV (the rest mass of the K-meson). Thus, with $\gamma^2 = 400$, $\Delta = 10^{-5}$ eV

$$\delta \ll \frac{\varepsilon \Delta}{4mV\gamma^2} \sim \frac{10^{-3} \times 10^{-5}}{10^3 \times 10^3} \sim 10^{-14},$$

using the mass of our galaxy. The choice of the galaxy is surely justified, because the range of the gravitational force certainly extends as far as a galaxy, since it is responsible for holding them together against the centrifugal force. If we dare to use the mass of the entire universe, we get $\delta \ll 10^{-20}$.

We conclude that, at least for kaons, the gravitational mass is strictly proportional to the inertial mass with remarkable accuracy.

III. Weakness

Looking at the numerical value of f one is struck by the large gap in orders of magnitude between the strength of gravitation and other known interactions.

GRAVITATION

If you compare, for example, the static electromagnetic force and the gravitational force, you find that the gravitational attraction between two protons is 0.8×10^{-36} less than the electric force:

$$f^2 m_p^2/r = 0.8 \times 10^{-36} e^2/r, \tag{3.1}$$

where f^2 is the gravitational constant, e the electric charge, and m_p the mass of the proton.†

We also define something which corresponds to the fine-structure constant α, namely the gravitational fine-structure constant α_G. Since $\alpha \sim 10^{-2}$, we find from Eqn (3.1) that

$$\frac{f^2 m_p^2}{\hbar c} = \alpha_G \simeq 0.5 \times 10^{-38}. \tag{3.2}$$

This is an exceedingly small number and therefore one sees, as a consequence, that on the elementary particle level the gravitational force is completely negligible. For instance, let us estimate the gravitational Bohr radius, that is the size of the lowest state of electron and proton attracted only by their gravitational interaction. Usually, the Bohr radius is given by:

$$r_B \sim \hbar^2/me^2 \sim 0.5 \times 10^{-8} \text{ cm}$$

m being the mass of the electron. The corresponding quantity for gravity r_{B_g} is

$$r_{B_g} = \frac{\hbar^2}{m^2 f^2 m_p} \sim 10^{31} \text{ cm}. \tag{3.3}$$

This is larger than the radius of the universe.‡

The same is true of lifetimes for the radiation of gravitons. This will be discussed in more detail later. We shall take a naive point of view for a rough estimate of the gravitational radiation. Assume that things work in a way similar to that of the electromagnetic case, except that now we have a coupling, not to the current, but to the energy–momentum tensor. Then one picks up an additional factor of the velocity v in the coupling constant and a factor v^2 in the emission rate.§ Therefore one finds for the gravitational lifetime, substituting α_G for α

$$\tau_G = \tau_{el} \cdot 10^{36} \cdot v^{-2}. \tag{3.4}$$

† This comes about in the following way, using c.g.s. units: $e \sim 10^{-10}$, $f^2 \sim 10^{-8}$, and $m_p \sim 10^{-24}$.

‡ This is estimated in the following way: Think of the universe as of something which is expanding with the speed of light for 10^{10} years; then its radius in c.g.s. units is
$$3 \times 10^{10} \times 10^{10} \times 10^7 \sim 10^{28} \text{ cm}.$$

§ One may also say that the additional v^2 arises because we have in this case quadrupole radiation.

Setting v equal to one (relativistic case) and taking for τ_{el} the shortest electromagnetic decay time, which is about 10^{-20}, we find:

$$\tau_G \sim 10^{18} \text{ sec.} \tag{3.5}$$

This is of the order of the lifetime of the universe; so one will hardly be able to wait long enough to see a graviton coming out of the decay of an excited nucleon state or the decay of elementary particles. Since α_G is so exceedingly small, one asks why one can see gravitation at all, because electromagnetic interactions are of the order of 10^{-2}, weak interactions are of the order of 10^{-5} (if we set the mass of the proton equal to one), and for the gravitational interaction we have suddenly this startling number 10^{-38}. First one might argue that for many particles gravitation always adds up whereas electric attractions will neutralize. However, one may object that second order effects in the electromagnetic interactions, like the van der Waals force due to the exchange of two photons do not neutralize and α^2 is still much larger than α_G. Also for the weak interactions there is an analogue of similar origin, the exchange of a neutrino–antineutrino pair. This should also give a long-range force, since particles with effective zero mass come into the game. However, the characteristics of such exchanges are that although in the case of two photons or two neutrinos particles with zero mass are available, zero mass would mean that they go parallel.† For this the phase space becomes zero, hence these forces decrease with a higher power of r than the Coulomb force, although not exponentially like the Yukawa force. Therefore, it turns out that over large distances, and in spite of the smallness of α_G, gravitation will dominate.

First, we take the familiar example of the van der Waals forces, where we get for the potential

$$V_{el} \simeq (e^2/r)(r_B/r)^5. \tag{3.6}$$

We have to compare this with the gravitational potential

$$V_G = \alpha_G/r. \tag{3.7}$$

From Eqns (3.6) and (3.7) we conclude that the condition that V_G should dominate over V_{el} gives

$$(r_B/r)^5 < 10^{-36}, \tag{3.8}$$

or that r has to be larger than $10^7 r_B$. Now since r_B is 10^{-8}, we find that already for a distance of $r = 0.1$ cm the gravitational force wins.

In the case of weak interactions the situation is somewhat questionable since higher order effects are divergent and the theory is not renormalizable. Nevertheless, in the following attempt to estimate the static potential, the

† The sum of two null-vectors with positive 0-component is again a null-vector if the space parts are parallel.

divergent aspects of the theory appear only in a very singular behaviour at $r = 0$. Since for us the region of large r is relevant, the divergence difficulties should not invalidate our conclusion.

If there is an electron–neutrino current, coupled to itself, one gets an interaction of the form

$$\mathscr{L}_I \sim (ev)(ev), \tag{3.9}$$

using the obvious notation. If there is a weak intermediary boson then you are bound to get terms of this form, which by a Fierz transformation (Marshak et al., 1969) can be written as

$$\mathscr{L}_I \sim (e\bar{e})(v\bar{v}), \tag{3.10}$$

i.e. an electron current coupled to a neutrino–antineutrino pair. Diagrammatically this appears as follows (Fig. 1):

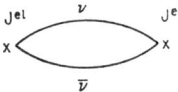

FIGURE 1. Neutrino pair exchange.

whilst in the electromagnetic system a photon is exchanged between two electric currents J^{el} (Fig 2):

FIGURE 2. Photon exchange.

It turns out that the difference between the weak force and the electromagnetic force is given by phase space. For the electromagnetic interaction one gets in momentum space $e^2\delta(p^2)$, where $\delta(p^2)$ is the effective mass distribution of the photon, whilst for the weak analogue we get G^2p^2, i.e. in all our calculations we have simply to substitute G^2p^2 for $e^2\delta(p^2)$ to get the results for the weak-interaction cases.† To calculate the force between two electrons, we have to remind ourselves that a particle with mass m gives a Yukawa-like force $\exp[-mr]/r$. To get the potential we have to average over the mass distribution, which one gets from phase space:

$$V_W \simeq G^2 \int dm^2 m^2 \exp[-mr]/r \simeq G^2/r^5. \tag{3.11}$$

† Dimensionally this is all right because e is dimensionless, $\delta(p^2)$ has the dimension $1/p^2$, which is $1/\text{mass}^2$ and G^2 is $(10^{-5})^2/m_p^4$.

We can write this in the following way:

$$V_W = 10^{-10} r^{-1} (\lambda_p/r)^4, \qquad (3.12)$$

where $\lambda_p = 1/m_p$ is the Compton wavelength of the proton. So we see that it does not decay quite as fast as in the electromagnetic case, but on the other hand we have a much smaller coefficient.

We now have to compare the gravitational potential V_G between two electrons and the weak potential V_W, and using Eqns (3.7) and (3.12) we find that for distances $r > 10^7 \lambda_p \sim 10^{-7}$ cm the gravitational force dominates over the force we get from weak interactions. Therefore, we see that for molecular distances the gravitational force will win over the weak forces and for macroscopic distances over the electric ones.

$$V_{\text{v.d.Waals}} = -\frac{10^{-2}}{r}\left(\frac{10^{-8}}{r}\right),$$

$$V_{\text{weak}} = -\frac{10^{-10}}{r}\left(\frac{10^{-14}}{r}\right)^4,$$

$$V_{\text{grav}} = -\frac{10^{-38}}{r},$$

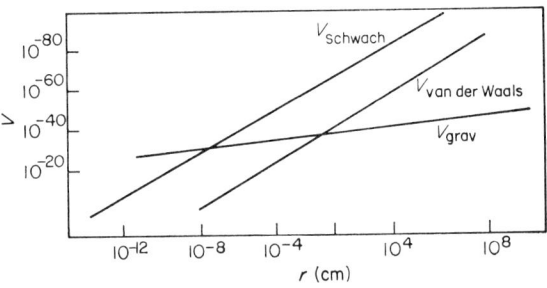

FIGURE 3. The potential of the weak, van der Waals and gravitational interaction as a function of distance (schwach = weak).

IV. Thermodynamic Behaviour of Gravitating Particles

Particles with gravitational interaction show a somewhat unusual (Lynden-Bell and Wood, 1968; Thirring, 1970) thermodynamic behaviour which we are going to discuss in this chapter. This behaviour is a consequence of the long range of the gravitational force and its universality which leads to the characteristic non-saturation of the gravitational energy. However, because of its weakness a large number of particles is required before these gravitational effects can manifest themselves. As mentioned in the introduction, clouds of gravitationally interacting matter have the tendency to contract and thereby to release energy and to become hotter. This somewhat perplexing affair is a direct consequence of the virial theorem which tells you that the total energy equals the negative kinetic energy, and therefore lowering the total energy means an increase in kinetic energy and therefore an increase in temperature. In this simple form the argument is certainly not quite correct because the virial theorem holds irrespective of the sign of the $1/r$ interaction. Since ordinary matter is dominated by Coulomb interactions, the argument would immediately also show that any piece of matter should have negative specific heat which we know to be false. First of all, one should not forget that in thermodynamics the system is always enclosed in a box and therefore there will be a contribution to the virial from the interactions with the wall; this external virial is directly related to the pressure. Secondly, the equipartition theorem according to which the kinetic energy is proportional to the temperature is no longer valid in quantum statistical mechanics since there is a contribution from the zero point motion to the kinetic energy. As a consequence, we shall see that at very high temperatures the contribution from the external virial and at very low temperatures the contribution from the zero point energy invalidate the argument and in the two limits the specific heat is positive. However, depending on other parameters, there may be a region in the middle where the specific heat is negative. This result may seem paradoxical and it happens, more strictly speaking, in the following way: In the microcanonical ensemble the specific heat† is related to the curvature of the entropy S as function E of the energy. Positive specific heat means that $S(E)$ is a concave function. If there is a region where $S(E)$ is convex, then the microcanonical and the canonical ensemble are no longer equivalent. In the canonical ensemble the specific heat is always positive and convex regions of $S(E)$ are bridged by a straight line corresponding to a phase transition. There is a simple physical reason for the fact that in the canonical ensemble where the system is in energy exchange with a large heat bath, the specific heat has to be positive. Indeed, a body with negative specific heat cannot be in thermal equilibrium with a large heat bath. The system may by a small fluctuation transfer energy to the heat bath and become hotter in doing so. Since the heat bath, being much larger, will heat up more slowly any temperature difference will be increased and no stable situation can be reached as long as the specific heat of the system is negative.

These features can actually be demonstrated rigorously since attractive $1/r$ potentials are perhaps the only realistic forces where the free energy for many fermions can be exactly calculated (Hertel and Thirring, 1971). Indeed, it

---

† The term heat capacity may here be more appropriate since it is a quantity referring to the system as a whole. It is not just the sum of the heat capacities of its parts because of the long range of the interaction.

turns out that in this case the temperature-dependent Thomas–Fermi equations become exact, as one might expect intuitively. The demonstration of the asymptotic exactness and the numerical solutions of the Thomas–Fermi equations involves considerable analytical and numerical effort. Nevertheless, the rough features can be guessed by rather simple reasoning as we shall show below. First of all, we have to show where for this system the usual proofs of the equivalence between microcanonical and canonical ensemble break down. They are based on a temperedness assumption, which means that the potential does not become too repulsive for large distances which certainly is satisfied for the Newton potential. The other assumption is that the ground state energy decreases proportional to N, the number of particles: this is where gravitating systems behave differently. Indeed, for fermions with with attractive $1/r$ interactions the ground state energy grows like $N^{7/3}$. This can be guessed by the following rough argument. If the system is confined within a region of dimension R the gravitational energy will be[†]

$$-\frac{\alpha_G N^2}{R}$$

In such a situation the volume available per fermion will be of the order $RN^{-1/3}$. Therefore, we get the kinetic energy owing to the zero point motion of the order $N(N^{2/3}/R^2)$. The sum of the two terms is minimized for $R \sim N^{-1/3}$ and the minimum value therefore proportional to $-N^{7/3}$.[‡]

Next we shall see what happens at finite temperatures. Although the well known expression of the thermodynamic quantities for free fermions involve the somewhat awkward Fermi function, we find that it is sufficient for our rough argument to interpolate between the high temperature limit where the system behaves like a classical ideal gas and the zero temperature limit where the zero point motion dominates. Therefore, we will construct the free energy per particle by adding to the zero point energy the free energy of the ideal Fermi gas and the potential energy corresponding to a uniform density in the volume V. It turns out that this procedure is numerically an acceptable approximation for the exact free energy of free fermions and the approximation of uniform density is not so bad because the system collapses altogether once the density becomes too inhomogeneous. In this way we get for the free energy per particle

$$\frac{F}{N} = \left(\frac{N}{V}\right)^{3/2} - T \ln \frac{V}{N} T^{3/2} - \frac{\alpha_G N}{V^{1/3}}. \tag{4.1}$$

[†] From now on we shall use natural units ($k = m = \hbar = c = 1$).

[‡] This simple consideration has been made rigorous by Lévy-Leblond (1970).

From this we can deduce by differentiation with the help of the standard thermodynamical relations† the following values for entropy per particle, energy per particle and pressure:

$$\frac{S}{N} = \ln\frac{V}{N}T^{3/2} + \frac{3}{2} = \frac{3}{2}\ln\left(\frac{V}{N}\right)^{2/3}\frac{2}{3}\left(E + \frac{\alpha_G N}{V^{1/3}} - \left(\frac{N}{V}\right)^{2/3}\right) + \frac{3}{2} \quad (4.2)$$

$$\frac{E}{N} = \left(\frac{N}{V}\right)^{2/3} + \frac{3}{2}T - \frac{\alpha_G N}{V^{1/3}} = E_K + E_P \quad (4.3)$$

$$P = \left(\frac{N}{V}\right)^{5/3}\frac{2}{3} + \frac{NT}{V} - \frac{\alpha_G N^2}{3V^{4/3}} = \frac{E_K + \frac{1}{2}E_P}{3V/2} = \frac{E - \frac{1}{2}E_P}{3V/2}. \quad (4.4)$$

The pressure has been rewritten in various ways to show that our approximation conforms with the virial theorem. Note that the pressure has three contributions. The thermal pressure $\sim V^{-1}$, the zero point pressure $\sim V^{-5/3}$, and the negative contribution due to the gravitational interaction $\sim V^{-4/3}$. Because of their different V-dependence the thermal pressure will dominate for large V and for small V the zero point pressure will win, but for sufficiently large N there will be a region in between where the pressure becomes negative. This is certainly physically impossible because the system is not glued to the walls but what will happen is that the system contracts by itself and thereby reaches a state of lower free energy.

Let us first consider the microcanonical case where the energy is the free variable. From Eqn (4.4) we see that for

$$E > -\frac{\alpha_G N}{2V^{1/3}} \quad (4.5)$$

† Remember that in the canonical ensemble, which corresponds to a system with Hamiltonian H in a heat bath with temperature T, the free energy is given by $F(T, V) = -T\ln\mathrm{Tr}\, e^{-H/T}$.

The other quantities are then

$$\frac{\partial F}{\partial V} = -P, \quad \frac{dF}{\partial T} = -S, \quad E = F + TS.$$

The microcanonical ensemble corresponds to an isolated system with energy E and there the entropy is defined by

$$S(E, V) = \ln\mathrm{Tr}\,\theta(E - H) \quad \theta(x) = \begin{cases} 1 & \text{for } x \geq 0 \\ 0 & \text{for } x < 0 \end{cases}$$

and we have

$$\frac{1}{T} = \frac{\partial S}{\partial E}, \quad P = T\frac{\partial S}{\partial V}.$$

Normally, the two ensembles give the same result but in our case this will not happen in the region of negative specific heat.

the pressure is positive whereas for

$$E < -\frac{\alpha_G N}{2V^{1/3}} \qquad (4.6)$$

it becomes formally negative. But this means that the system will contract to a volume V_0 such that

$$E = -\frac{\alpha_G N}{2V_0^{1/3}}. \qquad (4.7)$$

This volume has now also to be inserted in the other thermodynamic relations and the relation between energy and temperature

$$\tfrac{3}{2}T = \frac{E}{N} + \frac{\alpha_G N}{V^{1/3}} - \left(\frac{N}{V}\right)^{2/3} \qquad (4.8)$$

becomes by substituting V_0 for V

$$\tfrac{3}{2}T = -\frac{E}{N} - \frac{4E^2}{\alpha_G^2 N^{4/3}}. \qquad (4.9)$$

Thus, the temperature is a linear function of energy for high energies and for low energies a parabolic function as indicated in Fig. 4.

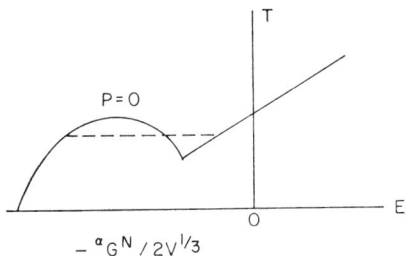

FIGURE 4.

One sees that actually the region with pressure zero begins with negative specific heat (T increasing with decreasing E), but at low energies the dominating zero point pressure makes the behaviour again normal.

Let us now consider the situation if we keep the temperature fixed and change the radius. The condition $P = 0$ now gives two solutions for the value of the volume. Since the pressure is proportional to the derivative of the free energy with respect to the volume, we see that the free energy is not monotonic in V. This implies that at a certain volume we can gain free energy by collapsing the system and thus the non-monotonic part of the curve will be bridged by a straight line. This corresponds to a phase transition where at constant temperature a finite amount of energy is released. This situation is illustrated in Fig. 5 and in Fig. 4 we have also drawn the corresponding pointed line which bridges the region of specific negative heat.

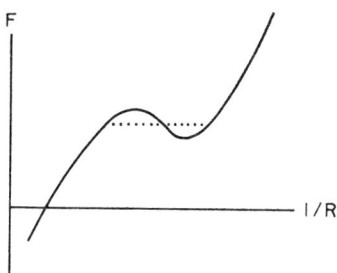

FIGURE 5.

After these considerations we will show in the following graphs the result of a computer solution of the exact Thomas–Fermi equations. For amusement we used for the mass of particles the mass of neutrons and the number

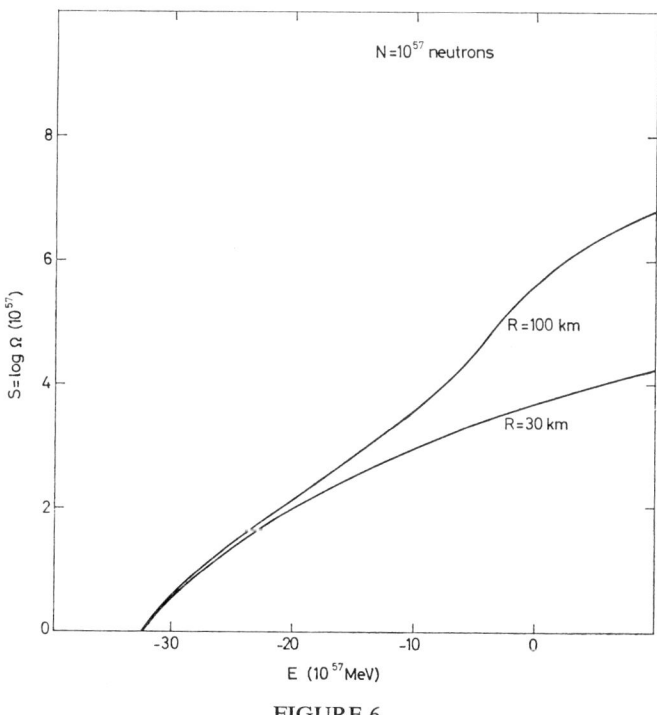

FIGURE 6.

N of particles corresponding to a star to illustrate that for this situation these phenomena happen at energies of several MeV per particle and of sizes for the system of several km. In Fig. 6 we plot the entropy as a function of the energy and show that for a radius of 100 km we have indeed a region where the function becomes convex. For the smaller radius this phenomenon is quenched by the zero point energy. The energy as the function of the inverse derivative of this curve, the temperature, is shown in Fig. 7. Thus, we have a region where we have three energy values corresponding to one temperature, and at a certain temperature the system in the canonical ensemble will jump from one branch of the curve to the other. The exact analysis verifies the usual statement that this happens when the free energy of the one branch becomes lower than the free energy of the other branch. In the next curve we plot the free energy as function of the temperature which clearly exhibits the point at which the free energy of one branch goes below the one of the other branch. The exact solution also gives the density as function of the radius and in Fig. 8 we plot the density for various temperatures. Note that above the transition point the density is reasonably homogeneous whereas at the phase transition the ratio between the density in the middle to the

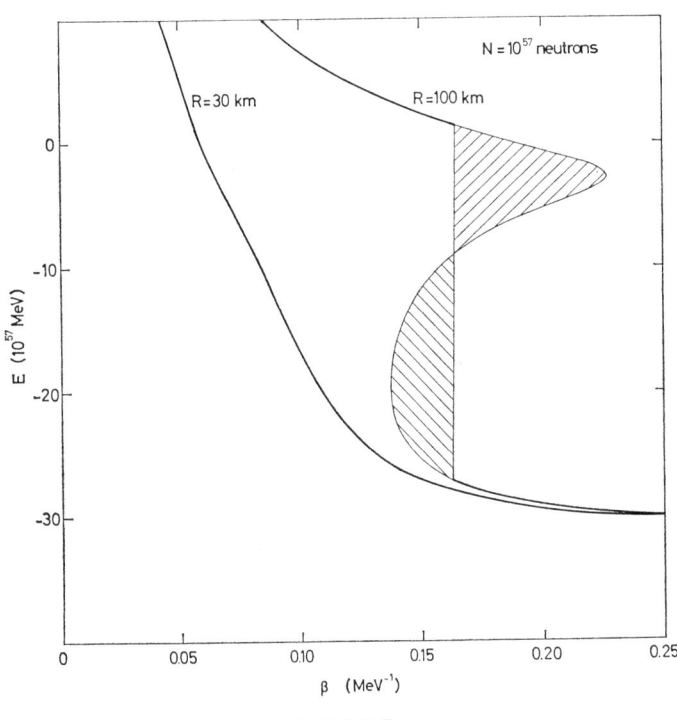

FIGURE 7.

density of the surface increases by about five orders of magnitude. Below the phase transition a kind of star develops with a reasonably well defined surface and a very thin atmosphere. This is also emphasized by a plot of the measure of the degeneracy of the Fermi gas (Fig. 9). The atmosphere always remains a Boltzmann whereas the centre becomes degenerate since it is only the zero point pressure which can stop the collapse.

The discussion given here shows the changes in thermodynamics that are going to happen if the gravitational interaction is included. Because of its long range the thermodynamic quantities do not have the usual extensivity property, and energy or entropy do not simply have twice the value of half the system. Because of this the usual thermodynamic arguments, indicating that specific heat and compressibility are positive, no longer hold. Indeed, there are regions where they become negative.

In the next Section we shall discuss how these results affect the life of stars and how they have to be changed to apply to realistic situations.

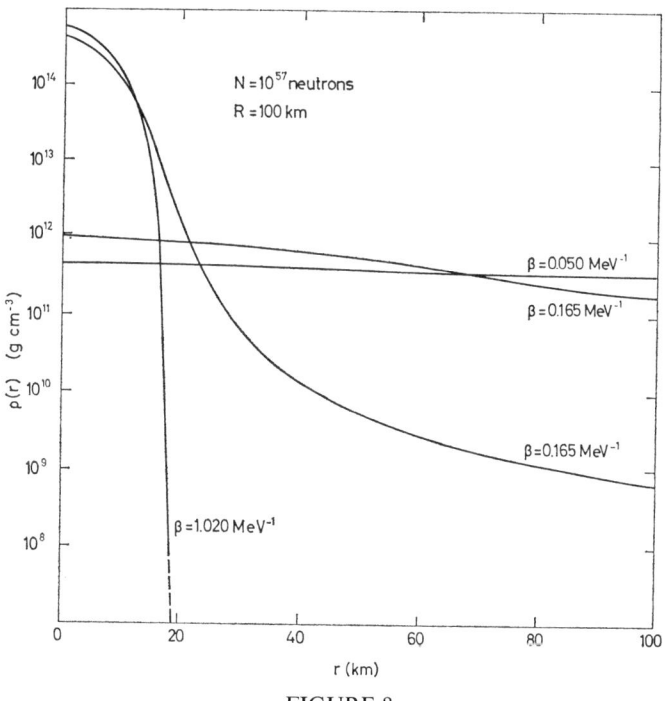

FIGURE 8.

V. Life and death of stars

If one wants to apply the considerations of the last chapter to stars then it is clear at once that the picture of a thermodynamic equilibrium will be only a rather rough model, since in a star the temperature is not constant and kinetic effects will be important. Furthermore, the chemical composition is not uniform and other factors like the radiation pressure may not be negligible (Chandrasekhar, 1939). Nevertheless, the equilibrium model reflects some essential features of such systems. Of course, now there is no box and therefore we have to consider the situation where the system is not constrained by an external pressure. This means that the virial theorem now applies without external virial. However, in the realistic situation it may happen that the particles which provide the zero point pressure (these are always the lightest particles, in ordinary matter the electrons) become relativistic. Also there are now nucleons and electrons, the former giving the main contribution to the gravitational interaction and the latter most of the zero-point pressure. In our units, with the mass of the proton = $h = c = k = 1$, we have now to introduce the mass of the electron m and we shall assume that there are as many electrons as protons. Then we get from the virial theorem the information:

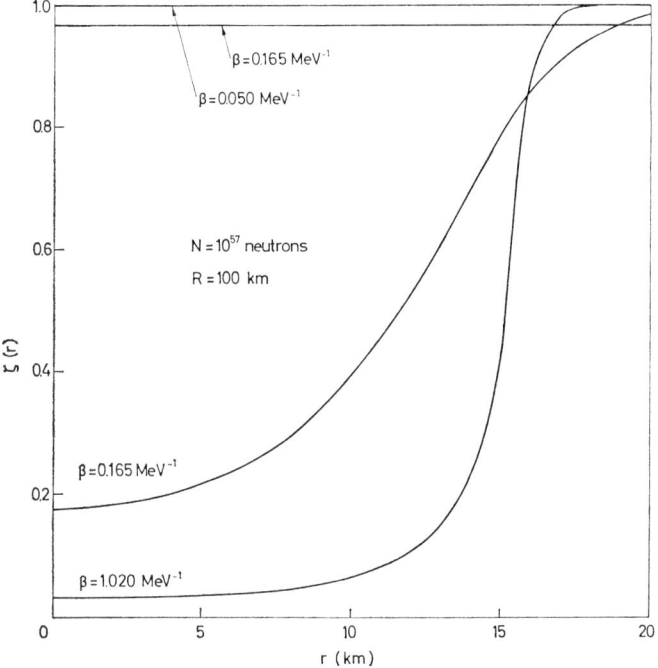

FIGURE 9.

GRAVITATION

kinetic energy per particle = thermal energy + zero-point energy

$$= \tfrac{3}{2}T + \sqrt{m^2 + \left(\frac{N^{1/3}}{R}\right)^2} - m$$

$$= -\tfrac{1}{2} \times \text{potential energy/particle}$$

$$= \frac{N\alpha_G}{R} = -\text{total energy/particle} \equiv -\varepsilon. \qquad (5.1)$$

These equations allow us to eliminate the radius R to which the system adjusts itself, and we get a relation between energy and temperature in the following form:

$$\tfrac{3}{2}T = m - \sqrt{m^2 + \frac{\varepsilon^2}{N^{1/3}\alpha_G{}^2}} - \varepsilon. \qquad (5.2)$$

A graph in which T is plotted as a function of ε (Fig. 10) immediately indicates that loss of energy leads again to a contraction of the system and to a rise in its temperature. If

$$N < N_c = (\alpha_G)^{-3/2} \sim 10^{57}, \qquad (5.3)$$

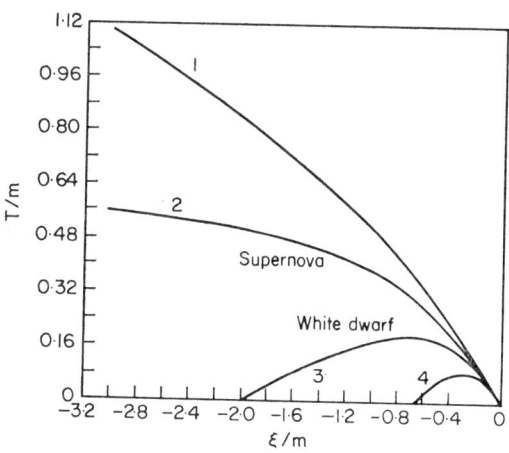

FIGURE 10. Star temperature versus the total energy per particle.

$$T/m = \frac{2}{3}\left[1 - \sqrt{1 + \left(\frac{N_L}{N}\right)^{4/3} \varepsilon^2/m^2} - \varepsilon/m\right].$$

(1) $\left(\dfrac{N_L}{N}\right)^{4/3} = \dfrac{1}{2}$;   (2) $\left(\dfrac{N_L}{N}\right)^{4/3} = 1$;   (3) $\left(\dfrac{N_L}{N}\right)^{4/2} = 2$;   (4) $\left(\dfrac{N_L}{N}\right)^{4/3} = 4$.

this process will continue up to a maximal temperature

$$T_{max} = \frac{2m}{3}\left\{1 - \sqrt{1 - \left(\frac{N}{N_c}\right)^{3/4}}\right\}, \tag{5.4}$$

which is reached for

$$RN^{-1/3} = \frac{1}{m}\sqrt{\left(\frac{N_c}{N}\right)^{4/3} - 1}, \tag{5.5}$$

and then the temperature decreases again.

The essential point is that the fateful number of particles, N_c, depends exclusively on the gravitational constant. If the system is to become a star, i.e. if the nuclear fuel is to be ignited, temperatures of (10–100) keV must be reached. Since, for electrons, $m/3 \sim 160$ keV, N must not be much smaller than N_c. For $N \sim N_c$, the electrons become relativistic ($RN^{-1/3} \sim 1/m$) and the energy of the electromagnetic radiation thus attains the order of magnitude of the energy in the matter, i.e. one particle with an energy of $\sim T \sim 1/RN^{-1/3}$ for each R^3/N. For $N \gg N_c$, the highly predominant radiation energy produces instabilities, so that visible stars can only be expected for $N \sim N_c$. Most stellar masses lie indeed in the $10^{33} g$ and $10^{34} g$ region, corresponding to $N \sim 10^{57}$.

For $N \gg N_c$, the relativistic electrons possess a zero-point energy of $\sim 1/RN^{-1/3}$. Consequently, if contraction continues, the increase of this energy no longer outweighs the lowering in the gravitational energy, and for $N > N_c$ the binding energy and the temperature will show an unbounded rise with decreasing $RN^{-1/3}$. Accordingly, in a sufficiently large star the zero-point pressure of the electrons will no longer be able to counterbalance the gravitational pressure and one of those cosmic catastrophes will occur which we see as supernovae. One can view this process as follows: Through the electromagnetic radiation the star is in heat contact with the rest of the universe which acts as heat bath. Therefore, the canonical ensemble should be considered and the system has, owing to its negative specific heat, the characteristic instability discussed in the last chapter. It will radiate energy and become hotter. Once the nuclear fuel ignites we stop in our way down the energy scale, because the energy radiated at the surface is supplied in the middle by nuclear energy. This pause continues until the nuclear fuel is burnt up and then the system becomes again hotter until new material can burn at the higher energy.† This unstable situation lasts as long as the specific heat is negative. For $N < N_c$ this state of affairs comes to an end and then the star radiates energy and cools down.

---

† This leads to the somewhat paradoxical conclusion that the function of the nuclear fuel is not to heat the star but to cool it since it prevents the star from heating up.

Finally it reaches the temperature of the background universe and thus becomes a black dwarf. For $N > N_c$ the situation never becomes stable and the star keeps contracting (its electrons become more and more energetic). If the energy of the electrons suffice for the inverse β-decay,

$$e^- + p \to n + \nu, \qquad (5.6)$$

a new and much more efficient mechanism for energy loss becomes available. Whereas photons need millions of years to get out from the middle of the star, neutrinos can leave almost instantaneously. Therefore our way down in energy is immensely speeded up and in a matter of seconds the electrons will be squeezed into the protons and the star collapses to a neutron star.

Certainly the detailed mechanism of this implosion involves a great deal of nuclear physics and kinetics. However, we can easily estimate the energy release and assume that a sizeable fraction of it ends up in electromagnetic radiation. Since it will be a few MeV per particle it is of the same order as the energy of the usual nuclear processes in stars except that now it is released much faster. If a supernova lasts a few days, we get the same energy during this time as from a normal star in 10^9 years which it needs to radiate its nuclear energy. Accordingly, the luminosity will increase by 10^{11} which is about the number of stars in a galaxy. This explains why in a supernova one star becomes as bright as a galaxy. For neutrons again our Eqn (5.1) is valid except that now the neutron mass instead of the electron mass has to be used. Since the gravitational and therefore the zero-point energy is of the order of a few MeV per particle the system has nuclear density and therefore a radius of about 10 km. The question arises whether strong interactions are in a position to prevent a further collapse (Wheeler, 1964). First of all, it is clear that the energy of even the strong interactions is not of a different order then the gravitational energy because both are of the order of the kinetic energy of the neutrons. Secondly, the critical number of particles does not depend on M, therefore it is not relevant whether we consider neutrons or electrons. Closer estimates of the influence of the nuclear forces show that they can push the deadly region away slightly and generate a stable situation at about $N_c < N < 2N_c$. However, for sufficiently large N gravitation will always win since its energy increases with a higher power of N than the energy of other interactions. Thus, for sufficiently large stars even the neutron star will not be the final state but the contraction will continue. If the system shrinks by another factor of 10 one has a radius of about 1 km and the zero-point energy and therefore the gravitational energy will increase by a factor 100 from 10 MeV to 1 GeV. This is just the rest energy of the neutron and then they will become relativistic. At the stage where the gravitational energy becomes equal to the rest energy our static considerations are no longer valid.† At these exceedingly strong gravitational fields new effects happen which can be described only in the framework of a relativistic field theory and we shall analyse them in the next Sections.

VI. Gravitational radiation

As a first effect which becomes important once the particles approach the speed of light we shall estimate the gravitational radiation assuming things work as in other field theories. There are two circumstances under which gravitation may become important. Either one has many particles or they are very energetic. We shall consider only the first possibility since the second one is beyond present experimental possibilities as gravitation becomes a strong interaction only for particles with a wave length of 10^{-33} cm. (Salam, 1970, 1971). As mentioned in the introduction gravitational radiation is similar to electric quadrupole radiation. There is no gravitational dipole moment, the latter is typical for electric interactions where you have positive and negative charges, whilst the gravitational charge has always the same sign. From electrodynamics we have the famous formula due to Larmor about the energy radiated in dipole radiation given an object with N charges e which rotates with a certain frequency ω. The energy loss per revolution is given by

$$\Delta E \text{ per } \omega = \omega \alpha N^2 v^2. \tag{6.1}$$

It goes with the square of the velocity v. For gravitational quadrupole radiation the analogue is (Landau and Lifshitz, 1959)

$$\Delta E \text{ per } \omega = \omega \alpha_G N^2 v^4. \tag{6.2}$$

Using $v = r\omega$ one gets the well-known r^2 dependence for the dipole and the r^4 dependence for the quadrupole radiation.

From weak interactions there is once again a competitor which should occur as a coherent process and therefore be considered. Whereas a first-order emission process for neutrinos is never coherent, for the loop (Fig. 1) one would have the same coherence as one would have for electromagnetic radiation, i.e. the amplitude of many particles can add together, and this in astrophysics is by no means a negligible effect. Just remembering our formula of Section III, this process behaves effectively as the emission of a particle with a continuous mass distribution, where the coupling is equal to the product of G^2 and m^2; one should then integrate the

---

† Remember that when the electrons became relativistic this meant only that the gravitational energy of the nucleons became of the order of the rest energy of the electrons.

contribution one gets from an individual mass with a factor $dm^2 m^2$, i.e.

$$G^2 \int dm^2 m^2 \delta(\omega - \sqrt{k^2 + m^2}) \tag{6.3}$$

(in electrodynamics it would just be $\alpha\delta(\omega - k)$). Thus one gets an additional factor ω^4, which one can also easily see using dimensional arguments:

$$\Delta E \text{ per } \omega \simeq \omega N^2 G^2 v^2 \omega^4 \sim \omega N^2 10^{-10} v^2 (\omega/m_p)^4. \tag{6.4}$$

Thus we find that this is exceedingly sensitive to the frequency; high frequencies like plasma oscillation may give you a very significant radiation of neutrinos. This can become a potent cooling mechanism in stars because the neutrinos can escape immediately.

To get a feeling for the gravitational radiation, we take a double star (radially collapsing objects do not give anything since there is no monopol radiation), and calculate the logarithmic decrement, i.e. the energy change per revolution divided by its energy E_G

$$\frac{\Delta E_G \text{ per } \omega}{E_G} \simeq \omega \alpha_G v^4 N^2 R / (N^2 \alpha_G) \simeq v^5. \tag{6.5}$$

We see that the strength of the interaction cancels out, so it is more or less a purely geometrical thing; for example, in the electromagnetic case one would get $\Delta E/E \sim v^3$. Because of the virial theorem, the gravitational energy is equal to the rotational energy mv^2 up to a factor of two. Thus, ΔE_G per ω goes effectively as v^7.

For planetary motion where $v \sim 10^{-4}$ the radiated power is negligible. However, when things become relativistic and $E_G \sim$ rest energy then $v \sim 1$ and a sizable fraction of the rest energy of a star may be radiated. Such violent situations may prevail in the collapse of a neutron star. Weber in his famous experiment has measured a flux of gravitational radiation of 10^4 erg/cm^2 s = 10^{12} m_p/cm^2 day. This corresponds to about one gravitational collapse per day in the centre of the galaxy, since it is at a distance $R \sim 10^{22}$ cm and $N/4\pi R^2 \sim 10^{57}/10 \cdot 10^{-44} \sim 10^{12}$. If this result is confirmed by other experiments now under way, it will further revolutionize astrophysics.

VII. The Equivalence Principle

In this Section we shall investigate why theories in which one has a universal coupling of a field to the energy momentum tensor of the form (1.6), satisfy a requirement stronger than universality: the equivalence principle. It can be formulated in the form that the effect of a constant external gravitational force can be compensated by a transition to a suitably chosen accelerated frame (Einstein, 1955). Clearly, if the effect of the gravitational field can be compensated by a transition to an accelerated frame the trajectories of particles will depend only on the initial conditions because in the new frame the gravitational field is transformed away so that we have the motion of free particles for which the statement is true. However, the converse is not true, universality does not imply equivalence. For instance, take electrodynamics where the charge as introduced in (1.4) is proportional to the mass of the corresponding particles and consider the motion of particles in a constant external magnetic field. The spiral motion of the particles will be the usual rotation with the Larmor frequency which is then the same for all particles, but the radius and axis of the spirals will be arbitrary. Therefore, there is no transition to one accelerated frame which transforms away the effect of the magnetic field for all particles at the same time.

Before discussing the effect of constant fields we first have to remark on the effect of constant potentials. The equations of motion ensuing from (1.5) or (1.6) are different from the ones coming from (1.4) in that even the values of the potentials and not only the field strength appear. This leads to some unusual effects and we first have to study their significance.

Consider the matter and interaction part of the Lagrangian as written earlier:

$$L_{\text{interaction}} + L_{\text{matter}} = \frac{m}{2}\int ds\, \dot{z}^i \dot{z}^k g_{ik}, \tag{7.1}$$

with

$$g_{ik} = \eta_{ik} + 2\psi_{ik}. \tag{7.2}$$

Solving the equations of motion that result from Eqn (7.1) leads to a geodesic in the space defined by the metric g_{ik}. But even without solving the equations one recognizes that the quantity

$$\dot{z}^i \dot{z}^k g_{ik} \tag{7.3}$$

is a first integral (constant of the motion), which therefore by a suitable choice of s can be normalized to 1. To appreciate the significance of this, consider electrodynamics where $g_{ik} = \eta_{ik}$. Then

$$\dot{z}^i \dot{z}^k \eta_{ik} = 1, \tag{7.4}$$

which can be rewritten in terms of its space and time components as

$$\left(\frac{dt}{ds}\right)^2 - \left(\frac{dx}{ds}\right)^2 = 1, \tag{7.5}$$

or equivalently in terms of the usual velocity

$$v^2 = \left(\frac{dx}{dt}\right)^2 = 1 - \left(\frac{ds}{dt}\right)^2 = 1 - \frac{1}{\gamma^2} \leqslant 1 \qquad (7.6)$$

since (ds/dt) is real. Equation (7.5) shows that the limiting velocity is 1, the velocity of light. On the other hand, the gravitational potential of the earth (1.6) inserted into (7.3) gives us

$$v^2 = \left(\frac{dx}{dt}\right)^2 = \frac{1}{1+2V}\left[1 - 2V - \left(\frac{ds}{dt}\right)^2\right] \leqslant \frac{1-2V}{1+2V}. \qquad (7.7)$$

Thus, the limiting velocity is different from 1 and is given by $\sqrt{(1-2V)/(1+2V)}$. For a static Newtonian potential, $V = kM/R$, so that the limiting velocity is less than the velocity of light. This, however, is not generally true since another choice of ψ_{ik} could lead to $v_{\text{limit}} > c$. This is a rather surprising result, because we have always the feeling that the velocity of light cannot be exceeded after all. We therefore ask: what is the velocity of a photon in the gravitational field? To calculate this you couple the gravitational and electromagnetic field in such a way that ψ_{ik} is multiplied by the energy–momentum tensor of Maxwell's theory. If one then solves Maxwell's equations, one finds that the field ψ_{ik} acts as a sort of dielectric constant and magnetic permeability, which changes the velocity of light to $\sqrt{(1-2V)/(1+2V)}$. Thus the photon has the same limiting velocity, but this limit is different from the velocity of light in free space. The theory is therefore reasonable in the sense that the photon acts like any other particle, except that the limit $m \to 0$ is taken. However, if the gravitational potential V is a function of space and time $V(x,t)$, so also will be the limiting velocity (that is, the velocity of light) $c' = c'(x,t)$.

We therefore recognize that there is no acausality in the sense that particles can go faster than the velocity of light. However, there is a peculiarity in that the velocity of light is a function of space and time.

We will now see that if the gravitational potential is constant, its effect can be eliminated by an appropriate linear transformation.

Note that in the presence of a potential of the form (2.3) the constant of the motion is

$$\dot{z}^i \dot{z}^k g_{ik} = (\dot{z}^0)^2(1-2V) - (\dot{z})^2(1+2V), \qquad (7.8)$$

which can be reduced to the free-field case

$$\dot{z}^i \dot{z}^k \eta_{ik} \qquad (7.9)$$

with the transformation

$$\left.\begin{array}{l} \bar{z}^0 = z^0 \sqrt{1-2V} \\ \bar{z}^{1,2,3} = z^{1,2,3} \sqrt{1+2V} \end{array}\right\}. \qquad (7.10)$$

However, if we have particles in a constant potential only, it is difficult to observe the change of scale indicated by Eqn (7.10) since practically nothing happens. Therefore, we look at a more instructive situation and assume that a Coulomb potential is present in addition to the gravitational potential. The motion of a charged particle in the presence of these fields tells us the influence of gravitation on the hydrogen atom. We shall indicate the derivation of the corresponding equations of motion in an appendix and state here only the result that the effect of the constant potential V can be included in the following change in the charge and mass of the particle:

$$e_{\text{eff}}^2 = e^2(1 - 2V), \qquad m_{\text{eff}} = m(1 + 3V) \qquad (7.11)$$

If we treat the hydrogen atom quantum-mechanically, we know that we will get stable solutions (Bohr orbits) with a Bohr radius

$$a_B = \frac{\hbar^2}{m_{\text{eff}} e_{\text{eff}}^2} \sim (1 - V)\frac{\hbar^2}{me^2} \qquad (7.12)$$

and a rotation frequency

$$\omega = \frac{c^2}{2\hbar^2} m_{\text{eff}}^2 (e_{\text{eff}}^2)^2 \sim (1 - V)\frac{c^2}{2\hbar^2} me^4. \qquad (7.13)$$

Hence, if the hydrogen atom is used as a unit, the lengths will shrink by a factor of $(1 - V)$, whereas the time will be lengthened by a factor of $1/(1 - V)$. This is exactly the meaning of the transformation given by Eqn (7.10). The observer located in the gravitational field who will measure the velocity of light in terms of Bohr radii per period of revolution, will find exactly the same result as the observer who is in a gravitationally field-free region. An "ideal" observer, on the other hand, who could measure the velocity of light in the gravitational field with his own "unscaled" rods and clocks would detect the change in the velocity, finding $c' = c(1 - 2V)$. Thus, a constant gravitational field does not produce observable effects. If the gravitational potential does not change too much over the dimensions of our measuring rods or clocks so that they are not too much affected by the gravitational force, the effect will be again the same as for the constant potential and result in a change of scale. However, since this change of scale will now be different at different points, there will be a global distortion of geometry. This is rather obvious and can be seen from the following trivial example. Consider the right-angled triangle with sides that are 3, 4 and 5 units long, shown in Fig. 11, and assume that a strong gravitational field exists in region G which shrinks the measuring rods there by a factor of, say, $\frac{1}{2}$. Then the measurement of the hypotenuse will give 10 units, so that even the triangular inequality will not be satisfied. Clearly this is an indication of a complete distortion of the geometry on a global scale.

It should be emphasized that the universality of gravitation is essential for its geometric interpretation. If the coupling constant f were different for different particles, then this would imply that measuring instruments made of the one or the other kind of particle are differently affected by the gravitational field and measurement with them would lead to different results. For example, it is well known that a strong magnetic field affects clocks, but the effect is not the same for all kinds of clocks. Watches made of steel are affected, watches made of plastic are not. Therefore, the effect of the magnetic field will not be interpreted as a change of space–time geometry. On the other hand, for gravitation there exists no instrument which is not affected and could measure the unscaled lengths and times so that an interpretation as a change of the geometry as defined by real measurments seems appropriate.

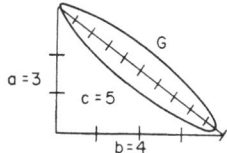

FIGURE 11. A local gravitational field G can change $c \to 10$.

We will now return to the case of a constant gravitational field. Again, for the theories we are considering, the effect of a constant field can be removed by a transformation which now, however, is quadratic. The classical limit for the motion in a constant force g is:

$$x = x_0 + gt^2/2 + v_0 t, \tag{7.14}$$

which can be transformed away by setting

$$\bar{x} = x - gt^2/2 \tag{7.15}$$

(that is, going to Einstein's free-falling elevator). For the general case we define the g_{ij} [see Eqn (7.1)] for a constant gravitational field

$$g_{ij} = \eta_{ij} + d_{il}^k \eta_{kj} z^l + \eta_{i\gamma} a_{js}^\gamma z^s + d_{il}^k z^l \eta_{k\gamma} a_{js}^\gamma z^s \tag{7.16}$$

where η_{ij} is as defined in the introduction. Note that (7.16) contains the general linear term but also a contribution quadratic in z^2. Without the latter the field can be compensated only up to order z^2. (Möller, 1952)

Now, what enters into the Lagrangian [Eqn (7.1)] is of the form

$$\dot{z}^i \dot{z}^j g_{ij}, \tag{7.17}$$

which reduces to the free-field form

$$\dot{\bar{z}}^i \dot{\bar{z}}^j \eta_{ij} \tag{7.18}$$

by the following transformations:

$$\bar{z}^i = z^i + \tfrac{1}{2} a^i_{kl} z^k z^l. \tag{7.19}$$

We therefore see that for the Lagrangian that we have chosen the principle of equivalence is incorporated, in the sense that the effect of a constant gravitational field can be eliminated by an appropriate (quadratic) transformation of the co-ordinates.

VIII. Strong and Weak Equivalence

We now need to go a step further in our theory, namely to add to the action the term W_{field} representing the gravitational field itself. That is, we allow the particles to create their own gravitational field, and we wish to study their gravitational interactions. It is now not clear whether our theory will still satisfy the equivalence principle, i.e. given a system of particles acting on themselves and placed in an external field, whether the effect of the external field can be eliminated by an appropriate transformation of co-ordinates.

We shall, therefore, distinguish between strong and weak equivalence, which we define as follows.

Weak equivalence: using the action $W_{\text{part}} + W_{\text{int}}$, the effect of an external field on a particle can be transformed away.

Strong equivalence: This corresponds in technical terminology to general covariance. In our language it implies that, using the action $W_{\text{part}} + W_{\text{int}} + W_{\text{field}}$, the effect of an external field can be transformed away also when the gravitational interaction between the particles is taken into account. For instance, a direct test of strong equivalence would be to measure the gravitational attraction between two bodies in a space ship far away from cosmic bodies and on earth in a freely falling elevator and compare the results. However, at present there is such direct experimental evidence only on weak equivalence and not on strong equivalence.

Some gravitational theories do satisfy strong equivalence and some do not, depending on W_{field}. An analysis of this point involves rather lengthy calculations because of the many possibilities for choosing a Lagrangian using a tensor field. We will not go through the algebra but will only indicate roughly the conclusions that can be reached (Sexl, 1967).

Consider the (symmetric†) tensor ψ_{ik}. ψ_{ik} has 10 components, and we can decompose it into the irreducible representations of the Poincaré group,

† The antisymmetric part will drop out anyway.

namely, into fields with specific spin. The decomposition gives

$$10 = 0 \oplus 0 \oplus 1 \oplus 2. \tag{8.1}$$

The first spin–0 part can be found by taking the trace $\psi^i{}_i$. The second spin–0 part is given by the double divergence $\psi^{ik}_{,ik}$. The spin–1 part is the divergence $\psi^{ik}_{,i}$ with the double divergence taken out, and the remainder is the spin–2 part of the tensor ψ_{ik}. The decomposition (8.1) can be checked by adding the dimensions of each representation

$$10 = 1 + 1 + 3 + 5. \tag{8.1a}$$

When the tensor ψ_{ik} is coupled to a conserved energy–momentum tensor, the second and third part of its decomposition, according to Eqn (8.1) the divergences $\psi^{ik}{}_{,i}$, do not contribute. To see this, consider the analogy with electrodynamics where we couple a conserved current j^i to a vector field A_i according to

$$A_i j^i.$$

Now the field A_i can be decomposed into a spin–1 part U_i that is divergence-less and a spin–0 part that is a pure gradient of a scalar: $\phi_{,i}$. If the current j_i is conserved, then the spin–0 part does not couple because

$$\phi_{,i} j^i = (\phi j^i)_{,j}$$

and a coupling term which is a pure divergence has no effect on the equations of motion.

The same thing happens with the tensor field ψ_{ik}. The two parts containing divergences, when coupled to a T^{ik} such that $T^{ik}_{,k} = 0$ do not affect the equations of motion. As pointed out by Gupta (1957) the conservation of T_{ik} requires that it also includes the contribution of the gravitation. Thus, one is automatically led to a self-coupling and therefore non-linear field equations.

Thus, if we use a conserved energy–momentum tensor, we are left only with the possibility of a spin–0 (scalar) field and a spin–2 (genuine tensor) field. Once the overall strength of gravity is fixed by the empirical gravitational constant, we are left only with the one free parameter δ. δ measures the ratio of the energy–momentum tensor of the spin–0 part to the spin–2 part of ψ_{ik}:

$$\delta = \text{percentage of spin–0}.$$

$\delta = 1$ corresponds to a pure spin–0 field, where

$$\psi_{ik} = \eta_{ik}\phi,$$

whilst $\delta = 0$ corresponds to a pure spin–2 field. Both the $\delta = 0$ and the

$\delta = 1$ case can be made to satisfy strong equivalence by the appropriate choice of L_f. (In the language of general relativity they can be made generally covariant.) On the other hand, theories that are admixtures of spin–0 and spin–2 parts ($\delta \neq 0, 1$) cannot, in general, be made to obey strong equivalence. Einstein's general theory of relativity corresponds to $\delta = 0$.

In the extreme case of $\delta = 1$ (spin–0 field), the g_{ik} would have the form

$$g_{ik} = \begin{pmatrix} 1 - 2V & & & \\ & -(1-2V) & & 0 \\ & 0 & -(1-2V) & \\ & & & -(1-2V) \end{pmatrix}, \qquad (8.2)$$

whereas for $\delta = 0$ one finds for the potential of a mass at large distances where the field is already weak ($V \ll 1$) the form

$$g_{ik} = \begin{pmatrix} 1 - 2V & & & \\ & -(1+2V) & & \\ & & -(1+2V) & \\ & & & -(1+2V) \end{pmatrix}, \qquad (8.3)$$

corresponding to (2.3). As seen by comparing (8.2) with Eqn (7.6), the clocks will slow down, but the measuring rods will now swell (instead of shrinking). Thus the velocity of light remains unchanged (on a global scale) and a g_{ik} of the form (8.2) will not affect the motion of light. This can also be seen because g_{ik}, as given above, will couple only to the trace of the energy momentum tensor $T^i{}_i$, which for the electromagnetic field is zero. Alternatively, we can consider the effective mass and effective charge picture that we developed before. Going through the same manipulations as in Eqns (A.1) to (A.10) (see Appendix), we find that the mass becomes lighter, $m_{\text{eff}} = m(1 - V)$ but, since there is no dielectric constant, the charge remains the same. Referring then to Eqns (7.12) and (7.13) we see that therefore the Bohr radius becomes larger, whereas the frequencies become smaller.

Since δ should be determined empirically let us then consider the three famous experimental tests of general relativity (Dicke, 1964) which now have reached an accuracy of about 1%.

(i) red-shift,
(ii) deflection of light,
(iii) motion of perihelion of Mercury.

Of these, the deflection of light already disproves a pure spin–0 theory. The deflection of a light ray in the vicinity of the sun is shown schematically in Fig 12. For a spin–2 theory the light ray stays away from the sun because a solution of Maxwell's equations leads to Fermat's principle. Since the closer to the sun the light ray passes the slower the velocity of light, the fastest path for the light ray will stay further away from the sun. Note that such a trajectory corresponds to an attraction between the photon and the sun. The dotted line shows the light ray for a pure spin–0 theory. The present evidence is consistent with $\delta = 0$.

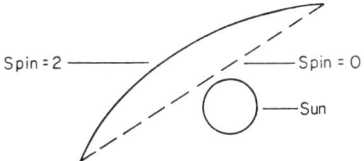

FIGURE 12. Deflection of light rays in the vicinity of the sun.

The red-shift does not differentiate between a spin–0 and a spin–2 theory. We can express the red-shift as showing only the slowing down of the clocks in a gravitational potential which, as we recall, is a common feature of both the g_{ik} of Eqn (8.2) and (8.3). However, it is good to keep in mind that we can look at the red-shift from two points of view: (a) In terms of the renormalized co-ordinates $\bar{z}$, in which case the atoms (by definition) maintain the right frequency. This is the usual way of presenting the red-shift, and then the photon loses frequency as it moves out of the gravitational field. Or (b), in terms of the "ideal" observer who sees everything in the unrenormalized co-ordinates. There the photon frequency does not change, but within the region of the gravitational field everything moves more slowly. Certainly the two things should not be added together.

Finally, the motion of the perihelion of Mercury is an extremely subtle effect. For its derivation we must take into account the coupling of the gravitational field to its own energy. In pure Einstein theory ($\delta = 0$) one gets exactly the observed value of 43″. However, since a small quadrupole moment of the sun may give corrections of the order of magnitude considered here, there would be room for a small admixture of spin–0.

One may ask why it is desirable to add a spin–0 field. The argument for such an addition is more of a philosophical than a scientific nature, and has to do with Mach's principle. Today's interpretation of it is, loosely speaking, that inertia is a consequence of the presence of other bodies in the universe (if only one body is present, then neither acceleration nor inertia have a meaning). Now Einstein claimed that his theory contained Mach's principle, because in a gravitational field, according to Eqn (A.9) (see Appendix), the effective mass becomes larger than the bare mass. Thus inertia increased due to the presence of the body that created the gravitational field.

The school of Dicke, on the other hand, points out that the effective mass effect is fictitious, since it is completely eliminated when the renormalized co-ordinates $\bar{z}$ are used. Thus, argues Dicke, Einstein's theory does not incorporate Mach's principle. As a matter of fact, Mach's principle is incompatible with strong equivalence, since to satisfy Mach, the Cavendish experiment must give a different result when performed in the presence of a third heavy body. Theoretically, this can be achieved by introducing a variable gravitational constant $K(x)$, which in turn corresponds to the addition of a spin–0 field ($\delta \neq 0$).

There exist several experimental proposals which will allow a more precise limit on δ.

IX. A Trip Behind the Schwarzschild Radius

The change in geometry which we discussed in Section VII is a completely negligible effect under usual circumstances. However, in the extreme situations which happen at a collapse of a neutron star, it becomes quite sizable and requires some rethinking of our intuitive notions. Again, reality may be rather complex but the relevant features appear already in an idealized situation which is described by the so-called Schwarzschild solution for the region outside a mass M. It is an exact solution of the rather complicated non-linear field equations and we just state the result in giving the corresponding g_{ik}.

$$ds^2 = dt^2 \left(1 - \frac{2M}{r}\right) - dr^2 \left(1 - \frac{2M}{r}\right)^{-1} - r^2 d\Omega^2 \qquad (9.1)$$

M is the mass of the object times the gravitational coupling constant and $d\Omega^2$ is the usual angular line element $d\Omega^2 = d\theta^2 + \sin^2\theta \, d\psi^2$. For $r \gg 2M$, (9.1) reduces to (8.3).

What is the physical meaning of r and t? In Section VII we showed that the influence of the gravitational potential can be viewed as a local scale transformation. All measuring rods and all clocks are affected in a certain way, in particular, the gravitational potential contracts rods and makes clocks go more slowly. The line element expresses just this effect. If we identify the dr and dt with the old co-ordinates which we first introduced and then found not to correspond to physical distances and physical times, then it is just ds^2 which we can measure with our clocks and our rods. And the factors $(1 - 2M/r)$, etc. tell you how much these quantities have changed. If you take a purely timelike distance you find that ds is smaller than dt, whereas d$r <$ ds for spacelike distances:

$$dr = d\Omega = 0, \qquad ds = dt\left(1 - \frac{2M}{r}\right)^{\frac{1}{2}} < dt,$$

$$dt = d\Omega = 0, \qquad |ds| \simeq dr\left(1 - \frac{2M}{r}\right)^{-\frac{1}{2}} > dr. \tag{9.2}$$

Thus, for purely spacelike distances the unit length, d$s = 1$, has length d$r < 1$ in unrenormalized co-ordinates, i.e. the measuring rod has shrunk. Analogously, one sees that clocks are slowed down.

Trouble arises when the factor $1 - 2M/r$ vanishes and you may ask what happens then. That something in the metric tensor becomes 0 is after all not so unfamiliar to us as it may seem at a first glance. In fact this happens even in polar co-ordinates where you have the usual $r^2 d\Omega^2$. This is equal to zero of $r = 0$ and this means that polar co-ordinates are not good for $r = 0$, but otherwise they are quite useful. If something becomes infinite as the radial component g in the Schwarzschild line element then this may seem something more dramatic and catastrophic. However, this may again just be due to the fault of the co-ordinates used and can happen in much more trivial cases. For instance, take a circle in the x–y plane and describe it always by, e.g., the x co-ordinate.

Then $y = \sqrt{1 - x^2}$ and

$$dy^2 = \frac{x^2}{1-x^2} dx^2, \qquad ds^2 = \frac{1}{1-x^2} dx^2. \tag{9.3}$$

So for $x = \pm 1$ it happens that $g = \infty$. In our previous language we could say that x corresponds to the unrenormalized co-ordinates and the ds are (Fig. 13) the renormalized real co-ordinates. Therefore the effect means bending the straight line into a circle and obviously equal distances in x will correspond to increasing distances in s and in particular at $x = 1$ the slope becomes

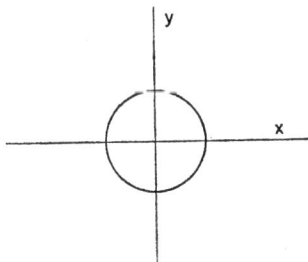

FIGURE 13.

infinite and we get into trouble. On the other hand we know that there is nothing particularly funny at these points. It is possible to introduce locally a co-ordinate system which corresponds to an Euclidean space. Obviously this is not necessarily globally true, as you can see with the example of the circle. Thus there are points where the old unrenormalized picture fails completely. Therefore one might ask whether our Schwarzschild manifold is just a piece of a larger manifold. And this turns out to be actually the case as can be seen by rewriting ds^2 in co-ordinates due to Kruskal (1969).†

The Kruskal co-ordinates are the following:

$$ds^2 = f^2(u, v)(dv^2 - du^2) - r^2(u, v)d\Omega^2,$$

$$u^2 - v^2 = \left(\frac{r}{2M} - 1\right)\exp(r/2M), \quad \frac{2uv}{u^2 + v^2} = tgh\frac{t}{2M}, \quad (9.4)$$

$$f^2(u, v) = \frac{32M^3}{r}\exp(-r/2M).$$

Motions in radial direction are described by

$$ds^2 = f^2(u, v)(dv^2 - du^2), \quad (9.5)$$

and for radial light rays the condition is

$$du^2 = dv^2 \quad (9.6)$$

and time-like lines are $|dv| > du$.

r = constant obviously corresponds to a hyperbola, and for t = constant u and v are proportional: $u = \alpha v$.

These co-ordinates are fine for $r \neq 0$; ds^2 becomes infinite just for $r = 0$, so we see that there is locally nothing particular going on at the curves $r = 2M$ which are the boundaries of the several regions in Fig. 14.

Let us see what this means physically. You can travel along lines which go steeper than 45°, so you may move in (I) and you may even cross the line $r = 2M$ rather easily even if these $t = \infty$. That is not absurd because the time is slowed down and it takes an infinite time t to reach the line $r = 2M$, the Schwarzschild radius. This is, however, measured in the original unrenormalized t co-ordinate which is what an observer measures at infinity. When measured in proper (renormalized) time it takes you a finite, even a rather short time to reach this line. Once you are in (IV) there is no escape and

---

† There are, of course, other ways of introducing co-ordinates and other ways of discussing it. For instance, there is a very nice embedding of this manifold in a pseudo-Euclidean space of higher dimensionality, which has been given by Fronsdal (1959).

eventually you will hit the line $r = 0$, that is to say you will sooner or later fall into the origin. Therefore a star which has collapsed through its Schwarzschild radius is called a black hole since from the outside one can notice only its gravitational field.

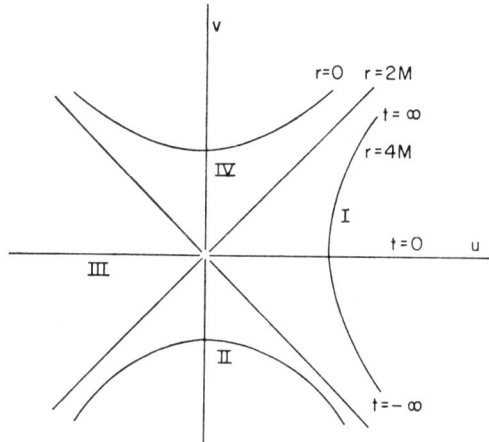

FIGURE 14. (I) $r > 2M$, $-\infty < t < \infty$; (II); (III); (IV).

There are two other regions, (II) and (III) which are new, (II) being the time-reversed region of (IV), all can go out, but nothing can come into (II).

From the outside we have no knowledge of (III), no light ray can connect (I) and (III).

For $r = 2M$ there is no particular singularity, nevertheless you may feel not very comfortable when you fall through the Schwarzschild radius, because the tidal forces can become terrific. What you really feel in a gravitational field is neither the potential nor the field strength but the gradient of the field strength, the tidal force. Because if everything is accelerated in the same way the field strength does not hurt you. But what does hurt you is a strong tidal force, that is to say the field down there is larger than up here, or in other words, it tends to pull you apart. The tidal force goes roughly like M/r^3; if $r = 2M$ the tidal force will go like $1/M^2$. For $M \sim$ solar mass macroscopic objects will be torn apart but for large objects like galaxies the force may not be too bad. Nevertheless, the line $r = 2M$ has some global significance, it is the point of no return. Furthermore, in going through the Schwarzschild radius you may be surprised in becoming aware of people whose existence was not known to you before. For instance, region (IV) is the only place where you can meet somebody from region (III). However, this situation is not specific for general relativity, it can even happen in special relativity: There it is possible that two fellows exist who cannot know of their mutual existence. Let the two hyperbolas of Fig. 15 be the world lines of these two fellows, and you can easily see that they will never succeed in communicating. So this sort of event horizon is nothing foreign to special relativity for accelerated motion, and here gravitation supplies the acceleration.

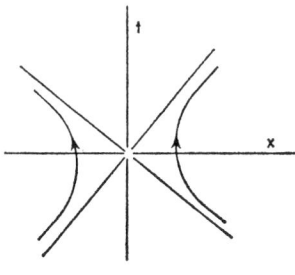

FIGURE 15.

The appearance of some infinities, e.g. $t = \infty$, is also not very disturbing. The proper time for all actual motions is finite. After some finite proper time one reaches $r = 0$ and we can show that there is no further enlargement of the manifold possible. The theory, therefore, does not tell you what happens then. But also this has its analogue happening in classical mechanics. A sufficiently strong repulsive potential, e.g. $\ddot{x} \sim x^3$ will push a particle after a finite time to infinity since

$$t_0 = \int_{x_0}^{\infty} \frac{dx}{\sqrt{E + x^3}} < \infty. \tag{9.7}$$

Classical mechanics refuses to make statements about $t > t_0$.

These and many other astonishing features are at present vigorously discussed in the literature (Thorne, 1967). One may argue that they belong rather to science fiction than to science because one will never see experimentally what happens inside the black hole. Nevertheless, these speculations are perhaps more than a philosophical amusement; firstly, because these things may happen in nature. Indeed, there has been recently a report (Cameron, 1971) of a "double star" where one partner, however, is invisible but must have a mass of about 23 solar masses. There is nothing reasonable we know which prevents objects of such a high mass eventually collapsing through the Schwarzschild radius and hence the interpretation of the invisible partner as a black hole seems to be the most conservative one. Secondly, because they teach us what radical changes of our usual notions happen once we consider couplings of the form (1.5), as innocent as they may look. If they occur between strongly interacting particles they will completely change our notions of space and time at small distances.

References

Cameron, A. G. W. (1971). *Nature Lond.* **229**, 178.
Chandrasekhar, S. (1939). "Stellar structure", University of Chicago Press, Chicago.
Dicke, R. H. (1964). "The Theoretical significance of experimental relativity", Gordon and Breach, New York.
Einstein, A. (1955). "The meaning of relativity", University Press. Princeton.
Eötvös, R. V., Pekar, D. and Fekete, E. (1922). *Ann. Physik* **68**, 11.
Fronsdal, C. (1959). *Phys. Rev.* **116**, 778
Good, M. L. (1961). *Phys. Rev.* **121**, 311
Gupta, S. N. (1957). *Rev. Mod. Phys.* **29**, 334.
Hertel, P. and Thirring, W. (1971). CMP 24 22. H. P. Dür (ed.) "Quointen und Felder", Braunschweig.
Kruskal, M. D. (1960). *Phys. Rev.* **119**, 1743.
Landau, L. and Lifshitz, E. (1959). "The classical theory of fields", Addison-Wesley, Reading.
Levy-Leblond, J. M. (1970). *Phys. Rev.* **D1**, 1837.
Lynden-Bell, D. and Wood, R. (1968). *Monthly Nat. Royal Astron Soc.* **138**, 495.
Marshak, R. E., Riazuddin and Ryan, C. P. (1969). "Theory of weak interactions in particle physics", Wiley-Interscience, New York.
Möller, C. (1952) "The theory of relativity", Clarendon Press, Oxford.
Particle Data Group. (1971). *Rev. Mod. Phys.* **43**.
Rohrlich, F. (1965). "Classical charged particles", Addison–Wesley, Reading.
Salam, A. (1970), 1971). ICTP-Trieste Preprints
Sexl, R. (1967). *Fortschr. Physik*, **15**, 269.
Schiff, L. I. (1959). *Proc. Natn. Acad. Sci. USA*, **45**, 69.
Thirring, W. (1970). *Z. Physik* **235**, 339.
Thorne, K. S. (1967). *Scient. A.* **217**, 88.
Weber, J. (1970). *Phys. Rev. Letters* **24**, 276.
Wheeler, J. (1964). "Gravitation and relativity", (H. Y. Chin ed.), Benjamin, New York.

Appendix

The equations of motion can be easily obtained (with some calculation), and read

$$m\ddot{z}^i g_{ik} = e\dot{z}^i F_{ik}, \quad (A.1)$$

which differs very little from the usual form. The quantity η_{ik} has simply been replaced by g_{ik}. On the right-hand side of Eqn (A.1) the current $e\dot{z}^i$ has been multiplied by the field tensor F_{ik}. In the case of a Coulomb field, only the 0th component will be different from zero and is given as the derivative of a potential A_μ which we write as

$$F_{0k} = A_{0,k}. \quad (A.2)$$

Thus the right-hand side of Eqn (A.1) reads $e\dot{z}^0 A_{0,k}$, where $A_{0,k} = ez/4\pi|z|^3$ in the absence of a gravitational field. The presence of the gravitational field, however, requires that we couple the electromagnetic field to it, with the result (as stated previously) of introducing an effective dielectric constant so that

$$A_{0,k} = \frac{ez}{4\pi|z|^3}(1 - 2V). \tag{A.3}$$

To approximate $\ddot{z}_0$ in the non-relativistic limit, we use $\dot{z}^i \dot{z}^k g_{ik} = 1$:

$$\left(\frac{dt}{ds}\right)^2 \left[(1 - 2V) - \left(\frac{dx}{dt}\right)^2 (1 + 2V)\right] = 1.$$

If $v^2 = (dx/dt)^2 \ll 1$, we drop the term in the velocity, so that

$$dt/ds \simeq 1 + V. \tag{A.4}$$

Hence the right-hand side of Eqn (A.1) reduces to

$$\frac{e^2 z}{4\pi|z|^3}(1 - V). \tag{A.5}$$

The left-hand side of Eqn (A.1) can be treated similarly:

$$\ddot{z}^{i=1,2,3} = \left(\frac{d^2 z}{dt^2}\right)\left(\frac{dt}{ds}\right)^2 \simeq \left(\frac{d^2 z}{dt^2}\right)(1 + 2V), \tag{A.6}$$

where we used Eqn (A.4); and finally the tensor g_{ik} for $k = 1, 2, 3$ is simply $(1 + 2V)$. Thus we obtain for Eqn (A.1) expressed in terms of the ordinary velocities and accelerations (in our non-relativistic approximation):

$$m\left(\frac{d^2 z}{dt^2}\right)(1 + 4V) = \frac{e^2 z}{4\pi|z|^3}(1 - V). \tag{A.7}$$

We can recast the result of Eqn (A.7) in the usual form for the equations of motion by introducing an effective electric charge and effective mass, which differ from the free charge and mass due to the presence of the gravitational field. For the electric charge we must write

$$e_{\text{eff}}^2 = e^2(1 - 2V), \tag{A.8}$$

since $(1 - 2V)$ is the effective dielectric constant, so that Eqn (A.7) will lead to Eqn (A.9) below:

$$m_{\text{eff}}\left(\frac{d^2 z}{dt^2}\right) = \frac{e_{\text{eff}}^2 z}{4\pi|z|^3}, \tag{A.9}$$

if we set

$$m_{\text{eff}} = m(1 + 3V). \tag{A.10}$$

The result given by Eqn (A.9) must also be obtainable from Eqn (A.7) if instead of introducing the effective charge and mass we use the change of scale indicated by Eqn (7.10). From Eqn (7.10) we obtain

$$\left.\begin{array}{l} \bar{z}^{1,2,3} \simeq z(1 + V) \\ \dfrac{d^2 \bar{z}^{1,2,3}}{dt^2} \simeq \left(\dfrac{d^2 z}{dt^2}\right)(1 + 3V) \end{array}\right\} \quad (A.11)$$

which when inverted and inserted into Eqn (A.7) gives

$$m\frac{d^2 \bar{z}}{dt^2} = \frac{e^2 \bar{z}}{4\pi |\bar{z}|^3} \quad (A.9a)$$

as expected.

Thus the effect of a locally constant gravitational field is equivalent only to a change of scale [Eqn (A.9a)], or we can say that according to Eqn (A.9) it is equivalent to a renormalization of the charge $e \to e_{\text{eff}} = e(1 - V)$ and of the mass $m \to m_{\text{eff}} = m(1 + 3V)$.